Efficiency of Vegetable Marketing
A Study with the Role of RMC

The Authors

Mr. Tushar Ranjan Sahoo, have qualification of UGC-NET, M.Phil., MBA, is a research scholar (PhD.) in Sambalpur University, Odisha, India. He has published three research articles in international journals, one research article in national journal and one chapter in a book. He has presented many research papers and cases in national and international conferences. He has two years corporate experience and six month research project experience.

Dr Saroj Kumar Sahoo, have qualification of PhD., UGC-NET, MBA, is working as Assistant Professor in PG. Departament of Business Administration, Sambalpur University, Odisha, India. He has published many research articles and cases in international & national journals. He has presented research papers and cases in more than twenty national and international conferences. He has thirteen years of teaching & research experience.

Efficiency of Vegetable Marketing
A Study with the Role of RMC

Tushar Ranjan Sahoo
Saroj Kumar Sahoo

2020
Associated Publishing Company®
A Division of
Astral International Pvt. Ltd.
New Delhi – 110 002

ISBN: 9789390371525 (Int. Edition)

Publisher's Note:

Every possible effort has been made to ensure that the information contained in this book is accurate at the time of going to press, and the publisher and author cannot accept responsibility for any errors or omissions, however caused. No responsibility for loss or damage occasioned to any person acting, or refraining from action, as a result of the material in this publication can be accepted by the editor, the publisher or the author. The Publisher is not associated with any product or vendor mentioned in the book. The contents of this work are intended to further general scientific research, understanding and discussion only. Readers should consult with a specialist where appropriate.

Every effort has been made to trace the owners of copyright material used in this book, if any. The author and the publisher will be grateful for any omission brought to their notice for acknowledgement in the future editions of the book.

Published by : **Associated Publishing Company®**
A Division of
Astral International Pvt. Ltd.
– ISO 9001:2015 Certified Company –
4736/23, Ansari Road, Darya Ganj
New Delhi-110 002
Ph. 011-43549197, 23278134
E-mail: info@astralint.com
Website: www.astralint.com

Acknowledgment

It is a great pleasure for me to acknowledge the assistance and contributions of many individuals in making this dissertation a success.

Special mention goes to my enthusiastic supervisor, Dr. Saroj Kumar Sahoo. My M.Phil. has been an amazing experience and I thank him wholeheartedly, not only for his tremendous academic support, but also for giving me so many wonderful opportunities.

Secondly, it is a pleasure to express my thanks to all respondents from interviews for sparing their time to participate in this study. I deeply appreciate their helpfulness and willingness in providing the useful information for this study.

Special thanks to all the faculty members and non-teaching staff of the department whose support and inspiration made this dissertation successful.

Lastly, I wish to express my sincere gratitude to my family for their encouragement and moral support.

Tushar Ranjan Sahoo
P. G. Department of Business Administration
Sambalpur University, Jyoti Vihar, Burla

Executive Summary

India is one of the leading producers of agriculture products but still, some peoples are starving for food with deep hunger. On the other hand, the producers of food, the farmers could not come out of the poor class family status and income level is so low that they are unable to maintain their lively hood. General consumers are suffering from the price hiking of the consumer goods. Therefore, a great question mark arises on marketing system of agriculture products in India. The marketing system will be more effective when the producers can directly sale their product to consumer. It is very much true, especially for the vegetable products. In this context the research problem is "Can the marketing system of vegetable product be efficient by integrating the consumers' positive orientation and farmers-cum-businessmen's happiness in future prospective?" From the above research question the aim of the study are to explore the factors of farmer-cum-businessmen satisfaction on vegetable marketing, to extract the factors of customers' satisfaction relating to vegetable market, to study the influence of factors farmers-cum-businessmen satisfaction on their happiness towards their current profession, to differentiate happiness of farmers-cum-businessmen over their demographic factor, to differentiate happiness of farmers-cum-businessmen over their marketing habits, to study the influence of factors of customer's satisfaction on consumers' positive orientation towards the vegetable shopping in RMC markets, to judge the interaction of demographic factors with the consumers' positive orientation towards the vegetable shopping in RMC market and to judge the interaction of consumers' purchasing habits with the consumers' positive orientation towards the vegetable shopping in RMC market. The current study will follow causative research design, where the relationships of satisfaction of farmers-cum-businessmen and the satisfaction of customers with the factors of marketing efficiency simultaneously are examined. The total sample size is 300 spreading over two set of questionnaires (150 from both farmers-cum-businessmen and consumers). Data is analysed by various descriptive statistics and empirical methods, such as scale reliability by Cronbach's alpha test, multiple linear regression method is followed to examine the degree of the farmer's satisfaction influencing the farmer's happiness and the degree of influence of consumers' satisfaction on vegetable consumers' positive orientation in the RMC markets. Famers' happiness and consumers' positive orientation integrally considered as 'marketing efficiency' in the present research. Three-way ANOVA technique is used to differentiate the farmers-cum-businessmen's happiness over the demographic factors of farmers-cum-businessmen and to differentiate farmers' happiness over the marketing habits of farmers-cum-businessmen's. The MANOVA technique is used to judge the interactions of demographic variables with the consumers' positive orientation (two dependent variables) in the RMC market and the interaction of purchasing habits with the consumers' positive orientations. The study concludes that it essential to maintain the present level of farmers-cum-businessmen's satisfaction and consumers positive thinking towards the vegetable marketing. There is a requirement of more and more number of market yards under RMC nearer to the farming area, by which farmers will not reluctant to come market yard. There should not to be bothering about the demographic factors of the farmers-cum-businessmen in the vegetable market because most of the time consumers will look towards the product benefits but not for the demographic factors of sellers as the demographic factors of consumers make variation on their product choice irrespective of farmers' demographic factors in the RMC market. So, the marketing strategies should be formulated by taking care of consumers' demographic factors only.

Contents

List of Tables

List of Figures

Plan of Study

Chapter - 1

- Chapter 1.1: Introduction
- Chapter 1.2 Industry profile
- Chapter 1.3: Impotence of study
- Chapter 1.4: problem statement
- Chapter 1.5: objective
- Chapter 1.6: Hypotheses

↓

Chapter 2.1 -REVIEW OF PREVIOUS RESEARCH FINDINGS
(Literature review -1)

- Chapter 2.1.1: Problem and prospect of vegetable marketing in India
- Chapter 2.1.2: Vegetable market Farmer-cum-businessmen prospective
- Chapter 2.1.3: Vegetable market consumer prospective

Chapter 2.2 - REVIEW OF THEORY
(Literature review -2)

- Chapter 2.2.1: Theory of agricultural marketing
- Chapter 2.2.2: Theory of agricultural marketing Indian context

↓

Chapter -2.3: RESEARCH GAP IDENTIFICATION

↓

Chapter □3: RESEARCH DESIGN AND METHODOLOGY

- Chapter 3.1: Designing
- Chapter 3.2: Sampling
- Chapter: 3.3: Respondent profile
- Chapter 3.4: Data collection method
- Chapter 3.4: Data analysis Procedure & techniques

↓

Chapter □4: DATA ANALYSIS AND INTERPRETATION

- 4.1: Descriptive statics
- 4.2: Empirical statistics

↓

Chapter -5 : SUMMERY FINDING

↓

Chapter-6: SUGGESITION /RECOMMENDATION

↓

Chapter □7: CONCLUSION

↓

Chapter -8: LIMITATION

↓

Chapter -9: ANNEXURE

- Chapter 9.1: Questioner
- Chapter 9.2: references
- Chapter 9.3: additional charts and graphs

CHAPTER - 1

Introduction

1.1 Introduction

India is abundant of natural resources but still,one-third of the populationis laying below the poverty line and most of the citizens are lower or middle-class people, which reflects an important economic issue that the uneven wealth distribution among the population. So, richer are becoming richer and poorer are becoming poorer. The needy poor people are struggling for two times of food and in another hand, the richer are wasting delicious foods. India is one of the leading producers of agriculture products but still, some peoples are starving for food with deep hunger. On the other hand, the producers of food, the farmers could not come out of the poor class family status and income level is so low that they are unable to maintain their lively hood. General consumers are suffering from the price hiking of the consumer goods. Therefore, a great question mark arises on marketing system of agriculture products in India. The marketing system will be more effective when the producers can directly sale their product to consumer.

A distribution channel is a chain of intermediaries through which products pass through till it reaches the end user. It includes stockist, wholesaler, retailer, distributor and even the online channels. The distribution channel is the path by which all the goods and services must travel to arrive the intending consumer. Conversely, it also used to describe the pathway of payment from the end users to producers. A distribution channel can be short or long depending up on the no of intermediaries present. Distribution channel of the fresh vegetable product has a very important role because this vegetable has a short life span and are not processed before delivery to the end user with fresh. Due to the perishability in nature and farmers have not a proper storage facility for the vegetable product; farmers have weak negotiation power as compare to the trader (Pokhel, D. M.,2010). In rural area, due to lack of infrastructure and transportation facility, the farmers are sales most their product to the intermediaries in state of direct to consumer(Mehta, and Sonawane, 2012). Farmers are using marketing channel as follows. (1). Producers–consumer (village sale),(2).Producer–retailer–consumer (local sale) (3).Producer–Trader–commission agent–retailer–consumer.(4).Producer–commission agent–retailer–consumer (4).Producer–primary wholesaler–secondary wholesaler-retailer– consumer (distant market). In Odisha, most of the vegetables are cultivated in rural area, but the farmer of the rural area are not getting justice because the price paid by the consumer mostly exploited by the traditional middle-men. So the farmers' (producers')share in the consumers' price is very less, which leads to the dissatisfaction in both ends (producers who get fewer returns for their cost & efforts and end-users who pay high price for the concerned products). The government machineries dealing with the agricultural product-marketing is rarely functioning for the fruit and vegetables. So most of the distribution channel of vegetables are dominated by the greedy land-lords / conventional business men. Most of the consumers' price is taken by the SAHUKARS (Land lords). Here the study access that the effectiveness of distribution channel, where channel effectiveness not means the benefit to only the intermediaries but also effectiveness of the channel for producer, consumer and intermediaries, so if the all channel holders including the consumers and farmers are benefitted from the channel the we can say that the channel is effective.

India is the world's second-largest producer of Fruits & Vegetables (F&V) with a total output of 227 million tons, and accounts for 15% of the global production of fruits and 10% of the global production of vegetables. The current F&V market in India is worth about INR 1.70 lakh crore which is expected to grow at a CAGR of 10% (Technopak, 2014). The above information shows that India is better position in fruits and vegetable which is enough for the self-sufficiency for the country's consumption and "Fruits and vegetables account for the largest portion of that wastage. 18% of India's fruit and vegetable production valued at Rs 13,300 crore is wasted annually," according to data compiled in a new report by Emerson Climate Technologies India, a business of the US-based manufacturing and technology company Emerson("India wastes Rs 44,000 cr of fruits, vegetables and grains annually | india | Hindustan Times," n.d.). On another end the needy people are not getting the vegetable in market at affordable price which shows that there is some difficulty in vegetable marketing.

As discussing the vegetable marketing, we can't ignore the medium through which we transport the product producer to end-user. In vegetable distribution system transportation plays a very important role in the distribution management. Without proper transportation the goods can't be delivered to the customer in a right time and in a right quality. It plays even a more important role in perishable food like vegetable because of short life, high perishable, required controlled temperature. Transportation related challenges are very high in the India because of unavailability of well transportation mode, high cost of transportation, lack of temperature controlled vehicle for the movement of goods etc. (Negi & Anand, 2015). So, to make distribution management there needed proper transportation mode form producer to consumer production of fruits and vegetable. In this context, this study intended to reveal the effectiveness RMC (regulated market commity) vegetable market yard through satisfaction of both the farmers-cum businessmen (farmers selling their product directly in RMC market yard) and satisfaction of consumer.

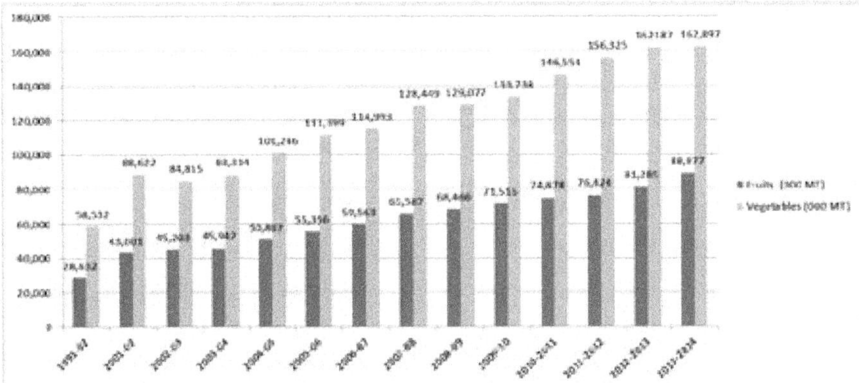

Fig. 1: Trends of Indian Fruits and Vegetable production

Source: - Indian Horticulture Database, NHB, 2015

As per the **Figure: 1,** there is constant increase in production of vegetable in India but still we are facing scarcity of vegetable in various parts of India. Till now, most of the small farmers are lying below poverty line; children in many places are facing malnutrition disease. In other hand vegetable productivity is increasing, which are rich of vitamin and minerals. Further, consumers are facing price hiking, even with the increased productivity that need to be affordable all the time by common middle class family. All of the above problem and discussion give a logical inference that lot of care and study need to be conducted for vegetable marketing by emphasizing satisfaction of both the consumer and the producer (farmer). In this context the research problem is stated in chapter-1.4

1.2 Industry profile

Vegetable find a special and important place in day-to-day life of Indians. India commands the world along various dimensions of agriculture, still lagging the developed countries' economy. A statistics worthy to mention here that 169.478 million metric tonnes of vegetables produced in the cultivated land of 9.542 million hectares (the data base of National Horticulture Board by the year 2014-15) which refers that India is ahead of most of the developed countries so far as vegetable products are concerned. By taking a specific data, it is known that India is the largest producer of Ginger and Okra amongst vegetables and ranks second in production of potatoes, onions, cauliflowers, brinjal, cabbages, etc. where India exports vegetables worth Rs.4,866.91crores in year 2015-16 as per the data base of National Horticulture Board by year 2015-2016. Another statistic by horticulture-statistical year book of India 2016 shows (table-1) that area of cultivation, production and productivity increases year-by-year. But most of the needy people are deprived of vegetables in India, which is reflected in the Global Hunger Index. India's rank is 97 in the index list of 118("Global hunger index: India ranks grim 97 of 118 countries on hunger index | India News - Times of India," 2016). So a vigorous study regarding above aspect is the need of the hour in the current nation's situation.

Table: 1 Trend of area wise production India and world

	India			World		
Year	Area(In Hectare)	Production (In Tonnes)	Productivity (In Tonnes/ Hectare)	Area(In Hectare)	Production (In Tonnes)	Productivity (In Tonnes/ Hectare)
2009-10	7984800	133737600	16.7	53709937	931085054	17.3
2010-11	8495000	146554000	17.3	53977066	1012524165	18.8
2011-12	8989541	156325481	17.4	58109601	1138121937	19.6
2012-13	9205186	162186567	17.6	58971121	1159179443	19.7

Source: Horticulture - statistical year book India, 2016

Above discussion shows that India is efficient in vegetable production but a lot need to be improved so far as storage and maintenance inefficiency is concerned as it leads to wastage of tons of food grains in the country. According to data compiled in a new report by Emerson Climate Technologies India, a business of the US-based manufacturing and technology company Emerson, about 18% of India's fruit and vegetable production valued at Rs 13,300 crore is wasted annually("India wastes Rs 44,000 cr of fruits, vegetables and grains annually | india | Hindustan Times," n.d.). Similarly un-uniformed decisions taken by decision maker or strategists at various stages of cropping and marketing of vegetable is causing continuous loss of farmers. Moreover, up-to-date information on prices and other marketing factors will enables farmers to negotiate with traders, facilitates spatial distribution of products from rural areas to towns & between the markets. It can also facilitate the farmer to produce right product at right time with right quantity so that farmers will not be harassed as the farmers are facing in the present condition of India. Fast transportation with minimum damage during shipment is very important in successful marketing of perishable products like vegetable. Therefore, efficient vegetable marketing embarks upon adequate storage, efficient transportation, reliable information and proper agriculture infrastructural facilities and appropriate planning and strategies for these aspects. Here, Njaya, (2014) stated that farmers are facing losses due to the poor infrastructure for storage, transportation and marketing facilities for fruits and vegetables of farmers and Presence of informal middlemen. Although the above dimensions are proven to be utmost requirements for the vegetable marketing in India, the planners and strategists could not do anything significantly for the country's economy. This can be attributed to the poor implemented policies of the marketing and distribution. This problem can be resolved by recognizing, understanding and considering the need and satisfaction of the both end participants (here, the agricultural producer and the consumers). So it is imperative to acknowledge the factors behind the overall satisfaction of both consumer and producer to make effective plans and policies relating to marketing efficiency on this context the research problem identified. (Referred in chapter -1.4)

1.3 Importance of study

- This study act as eye-opener for the government machineries those are dealing with distribution of vegetable directly or indirectly.
- It also helps to the consumer those who are facing problem like price-hiking, the low quality of the vegetable available in market and lack of information regarding pricing and quality of the product.
- The study also helpful to the organised channel members and deliver product to the needy consumer at affordable price with the good quality.
- This study also helpful to the farmers-cum-Businessmen to sustain in their profession and their better future.

1.4 Research Problem

Vegetable market effectiveness depends on various factors like transportation, warehousing, storage, and available technology, cost associate with these factor and government policies. Still there is no specific answer to make the agriculture marketing effective, especially of vegetable products. So, it needs special attention for its perishable characteristic. Further, the farmers in the Indian scenario are exploited by conventional businessmen, for which government has taken so many steps. One among them is regulated market committee (RMC) where the producers themselves supposed to be the businessmen. By this way end-users of vegetable product supposed to get the justice also. This aspect should ultimately reach the effective marketing system, which is rarely seen in India. In this context the research problem can be defined as **"Can the marketing system of vegetable product be efficient by integrating the consumers' positive orientation and farmers-cum-businessmen's happiness in future prospective?"**

1.5 Objective of the study:

1. To explore the factors of farmer-cum-businessmen satisfaction on vegetable marketing.
2. To extract the factors of customers' satisfaction relating to vegetable market.
3. To study the influence of factors farmers-cum-businessmen satisfaction on their happiness towards their current profession.
4. To differentiate happiness of farmers-cum-businessmen over their demographic factor.
5. To differentiate happiness of farmers-cum-businessmen over their marketing habits.
6. To study the influence of factors of customer's satisfaction on consumers' positive orientation towards the vegetable shopping in RMC markets.
7. To judge the interaction of demographic factors with the consumers' positive orientation towards the vegetable shopping in RMC market.

8. To judge the interaction of consumers' purchasing habits with the consumers' positive orientation towards the vegetable shopping in RMC market.

1.6 Hypotheses

H_1: There is a significance impact of factors of farmers' satisfaction on the farmers-cum-businessmen's happiness.

H_2: The customers' satisfaction in vegetable market is significantly influenced by consumers' positive orientation towards the vegetable market.

H_3: There is significant interaction of demographic factors of consumers with the consumers' positive orientation towards the vegetable shopping in RMC market.

H_4: There is significant interaction of consumers' purchase habits and the consumers' positive orientation towards the vegetable shopping in RMC market.

H_5: There is significant difference among happiness of farmers-cum-businessmen's and their demographic factors.

H_6: There is significant difference among happiness of farmers-cum-businessmen's and their marketing habits in vegetable market.

CHAPTER - 2

Literature Review

2.1 Review of previous research findings

2.1.1 Indian vegetable marketing problem and prospects

Vegetable is most essential diet in human nutrient, where sufficient production and proper distribution require for accesses the each and every citizen in the nation. According to the world summit on food security held on 16-18th Nov 2009 in Rome world population increase 50%, here the sufficiency of food for the people is not only the key challenge but also accessibility of healthy and proper food is main challenge of food security. In this context planner or strategist have to focus on proper distribution and production of food, where as vegetable has an important role in balance nutrient. In Indian scenario vegetable production is not a major issue in today's era. As per the data base of National Horticulture Board by the year 2014-15,India is the largest producer of Ginger and Okra amongst vegetables and ranks second in production of potatoes, onions, cauliflowers, brinjal, cabbages, etc. and overall second position in world in vegetable production. But still most of the Indian population deprive of proper food, which reflects in the hunger index India ranks 100 out of 118 countries (Mukherjee & Waghmare, 2017). All of the above evidence shows that there is some problem in vegetable marketing in India. In this context Deliya, Thakor, & Parmar(2011) found that their study in current unorganized retail scenario, the drawback is long distribution channels, which makes unorganized retailer to sale F&V to consumer at high price. The unorganized distribution channel has more wastage due to number of time load and unload and not having scientific handling and packaging. There is imbalance between the supply and demand due to poor forecasting, which boost the price in some months or seasons. And also the vegetables are not plucked from the firm due to the less demand. To supporting above statement M. B. Dastagiri et al.(2013) says that the study found marketing efficiency is highest in 'producer to consumer channel' as compare to other. In case most of the commodity, marketing cost, marketing margin, and transport cost, labour charges are adversely affecting marketing efficiency and open market price. All of above discussion shows that there is problem in distribution system but there is some how problem the nature of product as it is perishable in nature. So, B.M. Mehta And Madhuri Sonawane(2012) stated that Marketing cost of produces are mainly affected by the perishability of produces, breakage, spoilage, grading, transportation, storage, unfair and wasteful trade custom, Seasonal demand and supply which multiply the cost of produces that makes variation in price paid to farmer for their produce. So, it requires proper distribution with quality and safety major of the product. Therefore, for overall growth of the firm and agricultural development of India are the cold chain facility, fragmented supply chain, integration between partners, taxation issue and cost of packaging of material. Further, technology and technique, farmers' knowledge and awareness, quality and safety standards, processing value addition and market information etc. are the factors also affect the overall growth of agricultural development in India. (Negi & Anand, 2015). Again, Pokhrel (2010) due to perishability nature of firms' product and lack of proper storage, firms have weaker price negotiation as compare to trader & high profit margin taken by the trader due to the poor information of price margin as compare to the trader.

Traditional vegetable marketing is mostly practicing in India like nation, where intermediaries has important role in vegetable marketing. But the problem lies with about 60% of price hike without any value addition by the middle men and farmer getting 35-65 % from consumer rupees. Morden supply chain (Fab Mall, Subiksha, Trinethra, Namdhari Fresh, reliance Fresh) was found to be more effective than the traditional supplier and again found that higher total return and net return were found in co-operative retail format than the morden retail format.(Narasalagi & Hegade, 2013). In case of Thailand Supermarkets play important roles in connecting farmers to markets through direct procurement while improving cultivation practices of participating farmers in order to enhance quality of produce. A number of policy gaps still exist to protect the interests of the farmers in maximizing their benefits, and in the areas of participation between public and private sectors (Srimanee & Routray,2012).

Discussing above Indian vegetable market is suffering mainly due to lack of proper policy or rule and regulation of vegetable, in lack of advance technology in vegetable marketing and lack of proper infrastructure for vegetable marketing, where informal middle-men are taking advantage of said lacuna and make the monopoly in the vegetable market. To eradicate above problem government has established the RMC (regulated market committee) market where farmers can sale their product directly to the consumer. By which the vegetable distribution system can be more efficient.

2.1.2 Vegetable market Farmer-cum-businessmen prospective

2.1.2.1 Marketing channel of vegetable marketing

Discussing above section it can be conclude as the direct vegetable market is more efficient than the longer distribution channel. In this context Dey (2012) stated that undivided Andhara Pradesh government established RYTHU BAZAR (A place for direct transaction between the farmer and the consumer) farmers are less affected by the middlemen and the wastage of vegetable is less as compare to commonly used marketing channel and some improvement identified for the income generation of farmers like, facilitating the transportation for farmer, building a storage place and develop proper information system. Here, Deliya,

Thakor, & Parmar (2011) stated that in current unorganized retail scenario, the drawback is long distribution channels, which makes unorganized retailer to sale F&V to consumer at high price. The unorganized distribution channel has more wastage due to number of time load and unload and not having scientific handling and packaging. And another author said that marketing cost of produces are mainly affected by the perishability of produces, breakage, spoilage, grading, transportation, storage, unfair and wasteful trade custom seasonal demand and supply which multiply the cost of produces that makes variation in price paid to farmer for their produce(B.M. Mehta And Madhuri Sonawane, 2012). And another author said that share of producer in consumer price is high in case of the less number of the intermediaries.(Joshi, 2011). by Munhuweyi, K., Munhuweyi, U. L.and Sigge G.(2016) conduct a research on "Postharvest losses of cabbages from retail to consumer and the socio-economic and environmental impacts" conducted to quantify the incidence of postharvest losses of cabbage at retail purchase and during consumer simulated storage . Here researchers found that After seven days in storage, high incidence of postharvest losses occurred, ranging from 12 percent under cold storage to 46 percent under ambient conditions. Above study conclude that the waste of vegetable can happen with the proper technique used for preservation. By preservation technique the vegetable losses can be reduce some extent but can't be avoid. To solve the above issue there must require the direct marketing channel where the consumer can get the fresh vegetable from the farmers directly.

Martinez, S. W.(2016) did a research in USA vegetable market . By the study researcher found that Prices at direct sales outlets are lower than grocery store prices throughout the year and across the USA. Prices at direct sales outlets for some product/location/season combinations were higher than or comparable to supercentre prices. The above statement shows that direct marketing is not compulsory that always in the favour of consumer. Similarly, direct market also some time not beneficial for the farmer as per Silva, Dong, Mitchell, & Hendrickson (2015) direct marketing channel falling to achieve the goal because, insufficient in production, unwillingness of the customer to pay the price needed by the farmer to cover their cost. Again, they found that more broadly across market channels, it appears that perceptions of profitability influence a farmer's perception of their quality of life. As such, efforts to assist farmers to increase their profitability in market channels that also provide a higher quality of life would be beneficial and more education and training related to farm finances, thus allowing for more accurate assessment of the profitability of selling into different market channels as it relates to their specific perceptions of quality of life and profitability. So, for better and efficient vegetable marketing co-operation between the channel member must require. As Mesa, j. c. p., and Gómez, E.G.(2014) find that cooperation strategies are shown to have positive effects on performance (market creation, promotion, quality, training, joint supply purchases and research ventures). Moreover, the retail channel and market diversification are observed to have a positive effect on the relationship between cooperation and the supplier's performance.

2.1.2.2: Commission agents and organized market

(Aparna & Hanumanthaiah, 2013) found in their research that the net price received by the farmer in supermarket channel was more compared to traditional channel. The total marketing cost incurred by growers was more in traditional channels than supper market channel. The producers' share in consumers' rupee is higher for supermarket channel than traditional market channel. Again, another author stated same thing as above that the organized retailer usually gives higher price and make faster payment to farmer than traditional retailer. Organized retailers buying centre are closer to the firm-location which reduce the transportation charge for farmer and also in organized retail loading and unloading charge also bear by retailer (Rajkumar & Jacob, 2010). The above discussion shows that organized retiling is more beneficial for farmer and more efficient than the traditional market channel. In this context an researcher stated that about 60% of price hike without any value addition by the middle men and farmer getting 35-65 % from consumer rupees. Modern supply chain (Fab Mall, Subiksha, Trinethra, Namdhari Fresh, reliance Fresh) was found to be more effective than the cooperative and traditional supplier (Narasalagi & Hegade, 2013).

Above all discussion shows that there is problem in traditional marketing channel, by considering the study of Karim & Biswas (2016) the market intermediaries are categorized in to three groups 1) value added, 2) non value added, 3) necessarily value added activity. About 64% of activities are non-value activity because of redundant in nature of non-value added activity which leads to price rises up to double of producer margin. Producer and consumer are not getting proper justice due to the inadequate access of the market information and underdeveloped infrastructure. Another author got the similar result that The farmers share on consumer's rupees is low due to unreasonably high price fixed by middlemen which discourage farmers to increasing their marketable surplus (Pramanik & Prakash, 2010). Again similar result got by Mehta and Sonawane (2012) in their study that most of the farmers are using commission agent to sales their products, by which around 57% of price paid by the consumer are received by the commission agents. Beside direct market channel and the organised market channel the farmers can get more beneficial by making co-operation among the farmer. As they sale the product in bulk amount the transaction cost will minimize and farmers will be more beneficial which reflect on the study by Ruben, R., Boselie, D. & Lu, H.(2017) . Here they found that that trade-offs between higher production cum-investment costs (i.e. fixed investments, variable production costs and economies of scale) with expected savings in transaction costs (governance and opportunistic behaviour) are registered that could hinder contractual delivery.

2.1.2.3: Wastage of the product

Another main challenge face by the farmer in the vegetable market is wastage of the vegetable. Vegetable is highly perishable product, which need to special care for storage and transportation. There is huge wastage during the post-harvest storage and handling due to improper bagging without crating, lack of temperature controlled vehicles, no cold chain facilities for preserving the produce.

(Sheoran A, 2015) If consider the banana post-harvest value chain analysis in Zimbabwe year 2011-2012 the post-harvest losses around 24-27 %, these bulk of the losses occurred at farm level during handling and transportation. The major factors contributing to banana postharvest losses were: unreliable transport, poor communication and coordination between producers and processors; lack of or inefficient temperature management and poor sanitation. Another cause of losses of vegetable can be the improper demand and supply analysis of the vegetable. As per Deliya, Thakor, & Parmar(2011) there is imbalance between the supply and demand due to poor forecasting, which boost the price in some months or seasons. And also the vegetables are not plucked from the firm due to the less demand. As per Shukla , M. & Jharkharia, S (2013) the main interest is towards consumer satisfaction and revenue maximization where, post-harvest waste reduction being a secondary objective. Lack of demand forecasting, demand and supply mismatch, lesser integrated approach etc. are the major causes of the post- harvest losses. It shows that if special care will take to reduce post-harvest losses the profit of farmers can also be maximize.

2.1.2.4: Market facility and market information

Vegetables are highly perishable in nature, the market infrastructure and the information regarding the marketing of vegetable most important factors to know the efficiency of the vegetable marketing. Here an author stared that due to perishability nature of firms' product and lack of proper storage, firms have weaker price negotiation as compare to trader & high profit margin taken by the trader due to the poor information of price margin as compare to the trader (Pokhrel, 2010). Another author also stated the similar thing that farmers are facing losses due to the poor infrastructure for storage, transportation and marketing facilities for fruits and vegetables of farmers and p resence of informal middlemen are main cause of reduction of profit margin of farmers (Njaya, 2014). In this case Kumar (2012) says that there is moderate level problems of marketing vegetables and fruits in farmers' market with reference to freight charges, lack of proper grading, high carriage and other handling charges, exploitation of growers by market force, lack of proper quality control, long distance of market access, seasonal production, long marketing channel, delay payment, lack of cold storage place, advance sales agreement, inadequate post-harvest care, monopoly of middleman, bulkiness of products and low exports and there is low level problems of marketing vegetables and fruits in farmers' market with reference to irregular supply, primitive method of selling and price fixation, packing and loading problems, quality variation in production, packing of products. But they found that high level problems of marketing vegetables and fruits in farmers' market with reference to damage cost, intermediaries' exploitative practices, perishability of product, transportation cost and high storage cost. As stated above major problem is intermediaries' exploitative practice but still more than half of the fresh fruits and vegetables produced flows through the wholesalers. Movement and distribution of fresh fruits and vegetables from the farm to the retailers is dominated by wholesalers. Consumer preferences as well as the consumption patterns and styles are also changing due to easy access to information. Markets are adjusting themselves by responding to consumers' needs and wants (Man, Nawi, & Ismail, 2009). So for efficient vegetable marketing proper vegetable marketing good information system require so that transparency will be maintaining and farmer-cum-businessmen and consumer both will be benefitted. In this case Gunwant, Hussain, R.C, Faisal Ali, & Rana (2012) claim that in traditional vegetable marketing there is there is problem of the information regarding the market price. Where as cooperative play important role in equal distribution among the farmers as income disparity and The share of consumer rupee is highest in case of co-operative involved marketing (Niroja, Mamoru, & Yukio Muto, 2015). In co-operative system the farmers are forms a group and the sale their product mutually, so that they get the information regarding market easily. So for vegetable marketing information system takes a vital role.

ICT (information and communication technology) help the farmer how much to produce and when to produce and also help to decide the channel through which the produce flow and the major issue of distribution of vegetable is lack of integration among the channel members. Each participant in the chain act as an independent agent with low level of business relationship (Sazzad, 2014). Again, another author stated that to reduce post-harvest losses and deliver fresh vegetable to the end user there should needed the collaboration between the supply chain partners, ICT (information and communication technology) plays an important role for farmers to decide what to produce , what are the value should added after harvest, and which channel should be choose for the flow of produces (Sazzad, 2014). Geng, X.(2014) found the similar result that Chinese aquatic farmers' relationship networks can positively impact their participation in modern channels through the mediators of trust, specific assets investments and joining in cooperatives.

Farmers' knowledge and awareness regarding the vegetable marketing is most essential for the efficient vegetable marketing. But sharing the knowledge among to other in same profession will be more fruitful. In this case Chipungahelo (2015) conducted a study in kilosa district of Tanzania to examine knowledge sharing on traditional vegetables for supporting food security among farmers and other communities in Kilosa district and researcher found that that farmers used a socialization approach to share indigenous knowledge about traditional vegetables on production, consumption and preservation and getting the fruitful result in vegetable marketing. Rajagopal and Bansal (2015) also stated that market awareness has an implant role in the vegetable marketing. From the above discussion it can be interpret that the for the efficient vegetable marketing there most require the good market infrastructure and proper communication and sharing of knowledge about the market.

2.1.2.5: Rules and regulation in vegetable marketing

For any of the organizational success there most require proper strategy and policy require. Likewise, for proper functioning of vegetable market it also needed the proper strategy and policy. In this context Gunwant, Hussain, R.C, Faisal Ali,

& Rana(2012) said that in Indian there is lack farmer association for advocacy, lack of policy regarding price fixation and crop insurance. So farmers are not getting their right which they deserve. Srimanee and Routray (2012) conducted a study on "The fruit and vegetable marketing chains in Thailand: Policy impacts and implications" by the study they found that for efficient vegetable marketing there needed to make policy of vegetable marketing in keeping view of both the consumer and farmer-cum-businessmen.

Policy making is not only the main issue in the Indian context but main issue is to implementing the policy. Properly implementation of the policy can make the vegetable marketing more efficient. Wongprawmas, R., Canavari, M. & Waisarayutt, C.(2015) conducted a study on "A multi stakeholder perspective on the adoption of good agricultural practices in the Thai fresh produce industry" . Here, researcher found that Stakeholders perceived the credibility of national GAP (Q-GAP) standards implemented in the Thai domestic market as low, due to a lack of reliable verification and traceability and because the system is managed by governmental bodies. Consumer demand could drive the market of safe produce, but retail stores are the main influencers of producers' adoption of GAPs in domestic markets. So, there is needed concrete rule and regulation on the policy adopted in the vegetable market which can be fruitful for both the consumer and the farmer-cum-businessmen.

2.1.3 Vegetable marketing in the consumer prospective

2.1.3.1: Price factor of consumer

In Indian vegetable market the consumers are mainly poor and middle class peoples. So, in vegetable market price matter more. So, little bit change in price of vegetable the entire vegetable market may affect. Whereas in vegetable market most of the businessmen are unorganized retailers and for them there is not such rule and regulation for vegetable marketing vegetable. Here Deliya, Thakor, & Parmar, (2011) stated that the current unorganized retail scenario, the drawback is long distribution channels, which makes unorganized retailer to sale F&V to consumer at high price and there is imbalance between the supply and demand due to poor forecasting, which boost the price in some months or seasons. And also the vegetables are not plucked from the firm due to the less demand. In this context another author found the similar result that most of the farmers are using commission agent to sales their products, by which around 57% of price paid by the consumer are received by the commission agents. Marketing cost of produces are mainly affected by the perishability of produces, breakage, spoilage, grading, transportation, storage, unfair and wasteful trade custom easonal demand and supply which multiply the cost of produces that makes variation in price paid to farmer for their produce. (B.M. Mehta and Madhuri Sonawane,2012). Here Karim & Biswas (2016) also stated that market intermediaries are categorized in to three groups 1) value added, 2) non value added, 3) necessarily value added activity. About 64% of activities are non-value activity because of redundant in nature of non-value added activity which leads to price rises up to double of producer margin. Producer and consumer are not getting proper justice due to the inadequate access of the market information and underdeveloped infrastructure (Karim & Biswas, 2016).

All of the above statement reflects that longer the channel vegetable marketing channel more price increase and product quality also declining gradually. So, here Dastagiri et al.(2013) stated that marketing efficiency is highest in 'producer to consumer channel' as compare to other. In case most of the commodity, marketing cost, marketing margin, and transport cost, labour charges are adversely affecting marketing efficiency and open market price. Same thing also said by Joshi (2011) share of producer in consumer price is high in case of the less number of the intermediaries. Supporting to above author Martinez (2016) found that Prices at direct sales outlets are lower than grocery store prices throughout the year and across the USA. From the above discussion we found that for satisfying the Indian consumer in vegetable market there needed to take care of both the price of the product, quality of the product and service provided by the seller.

2.1.3.2 Marketing support in vegetable marketing

Only price doesn't matter everywhere, there also good service and the quality product also matter in some extent in the market. In this case Wongprawmas, Canavari, & Waisarayutt, C. (2015) found that Stakeholders perceived the credibility of national good agricultural practices (GAP) standards implemented in the Thai domestic market as low, due to a lack of reliable verification and traceability and because the system is managed by governmental bodies. Consumer demand could drive the market of safe produce, but retail stores are the main influencers of producers' adoption of GAPs in domestic markets. Here Nandonde & Kuada(2016) stated that the criteria used by modern food retailers in the selection of local food suppliers are reliability, quality, trade credit and legal certification. Again, Gyau and Spiller (2009) said that in the fresh produce business, the level of perceived power exploitation may differ depending on the type of business organization or the position of the firm in the supply chain. Here, Pe´neau, Linke, Escher & Nuessli (2009) is also stated that freshness of vegetable can be derived from the level of closeness to the original product, in terms of distance, time and processing. After-all the main intention of the vegetable marketing is satisfying the consumer and revenue maximization. Here Shukla & Jharkharia (2013) stated that main interest is towards consumer satisfaction and revenue maximization and other activities are the secondary objective.

So from all of the above discussion it can be said that the vegetable marketing will be efficient if both the consumer and the farmer-cum-businessmen are satisfied with their role. So here the policy maker should make decision keeping the view point of both the consumer and farmer-cum-businessmen.

2.2 - Review of Theory

2.2.1Agricultural Marketing:

Agriculture means activities aimed at the use of natural resources for human welfare, that means it includes all the primary activities of production. But, generally, it is used to mean production crops and livestock. Marketing refers to a series of activities involved in moving the goods or services from the producer to consumer. It includes all the activities involved in the creation of time, place, form and possession utility. According to Thomsen, the study of agricultural marketing, comprises all the operations, and the agencies conducting them, involved in the movement of farm-produced foods, raw materials and their derivatives (Kahlon And Tyagi,2006).

2.2.1.1: Nature of vegetable marketing

Perish-ability of the Product: Most of vegetable and related products are perishable in nature; but the period of their perishability varies from a few hours to a few months.

Seasonality of Production: Farm products are produced in a particular season; they cannot be produced throughout the year. In the harvest season, prices fall.

Bulkiness of Products: The characteristic of bulkiness of most farm products makes their transportation and storage difficult and expensive. This fact also restricts the location of production to somewhere near the place of consumption or processing. The price spread in bulky products is higher because of the higher costs of transportation and storage.

Variation in Quality of Products: There is a large variation in the quality of agricultural products, which makes their grading and standardization somewhat difficult. There is no such problem in manufactured goods, for they are products of uniform quality.

Irregular Supply of Agricultural Products: The supply of agricultural products is uncertain and irregular because of the dependence of agricultural production on natural conditions. With the varying supply, the demand remaining almost constant, the prices of agricultural products fluctuate substantially.

Small Size of Holdings and Scattered Production: Farm products are produced throughout the length and breadth of the country and most of the producers are of small size. This makes the estimation of supply difficult and creates problems in marketing.

Processing: Most of the farm products have to be processed before their consumption by the ultimate consumers. This processing function increases the price spread of agricultural commodities (Kulkarni, 1964).

2.2.2: Market functionaries:

In the marketing of agricultural commodities, the following market functionary's/ marketing agencies are involved:

Producers: Most farmers or producers, perform one or more marketing functions. They sell the surplus either in the village or in the market. Some farmers, especially the large ones, assemble the produce of small farmers, transport it to the nearby market, sell it there and make a profit. This activity helps these farmers to supplement their incomes.

Middlemen: Middlemen are those individuals or business concerns which specialize in performing the various marketing functions and rendering such services as are involved in the marketing of goods. The middlemen in food grains marketing may, therefore,

Wholesalers: Wholesalers are those merchant middlemen who buy and sell food grains in large quantities. They may buy either directly from farmers or from other wholesalers.

Retailers: Retailers buy goods from wholesalers and sell them to the consumers in small quantities. They are producers personal representatives to consumers. Retailers are the closest to consumers in the marketing channel.

*Itinerant Traders***:** Itinerant traders are petty merchants who move from village to village, and directly purchase the produce from the cultivators. They transport it to the nearby primary or secondary market and sell it there.

Village Merchants: Village merchants have their small establishments in villages. They purchase the produce of those farmers who have either taken finance from them or those who are not able to go to the market. Village merchants also supply essential consumption goods to the farmers. They act as financers of poor farmers. They often visit nearby markets and keep in touch with the prevailing prices.

Survey and research studies and in providing guidance, technical and financial support to the State Governments (maoria and joshi, 1995)

2.2.3 Regulated market

Under the traditional system of marketing of the agricultural products, producer sellers incurred a high marketing cost, and suffered from unauthorized deductions of marketing charges and the prevalence of various malpractices. To improve marketing conditions and with a view to creating fair competitive conditions, the increase in the bargaining power of producer-sellers was considered to be the most important prerequisite of orderly marketing. Most of the defects and malpractices under, the then existing marketing system of agricultural products have been more or less removed by the exercise of public control over markets, i.e., by the establishment of regulated markets in country.

Regulated Market Committee is one which aims at the elimination of the unhealthy and unscrupulous practices, reducing marketing charges and providing facilities to producer-sellers in the market. Any legislative measure designed to regulate the marketing of agricultural produce in order to establish, improve and enforce standard marketing practices and charges may be termed as one which aims at the establishment of regulated markets. Regulated markets have been established by State Governments and rules and regulations have been framed for the conduct of their business. The establishment of regulated market is not intended at creating an alternative marketing system.

The basic objective has been to create conditions for efficient performance of the private trade, through facilitating free and informal competition. In regulated markets, the farmer is able to sell his marketed surplus in the presence of several buyers through open and competitive bidding. The legislation for the establishment of regulated markets does not make it compulsory for the farmer to sell his produce in the regulated market make it compulsory for the farmer to sell his produce in the regulated market yard. The basic philosophy of the establishment regulated markets is the elimination of malpractices in the system (Acharya and Agrawal, 2006).

2.2.3.1: The specific objectives of regulated markets:

1. To prevent the exploitation of farmers by overcoming the handicaps in the marketing of their products,
2. To make the marketing system most effective and efficient so that farmers may get better prices for their produce and the goods are made available to consumers at reasonable prices.
3. To provide incentive prices to farmers for inducing them to increase the production both in quantitative and qualitative terms,
4. To promote an orderly marketing of agricultural produce by improving the infrastructural facilities.

2.2.3.2 Importance of regulated markets

Market area: The area from which the produce naturally and abundantly flows to a commercial centre, i.e., the market, and which assures adequate business and income to the market committee

Principle assembling market: It is the main market which is declared as a principal market yard on the basis of transactions and income generated for the market committee

Sub market yard: It is sub yard of the principle assembling market. This is a small market and does not generate sufficient income to declare as a principal assembling market

Market yard: This is a specified portion of the market area where the sale, purchase, storage and processing of any of the specified agricultural commodities are carried out (Acharya and Agrawal, 2006).

2.2.3.3: The ways Market Committee functions:

i) Maintain and manage the market yards and sub-market yards within the market area.

ii) Provide the necessary facilities for the marketing of agricultural produce within the market yards and outside the market yards and within the sub-market yards and outside the sub-market yards in the market area;

iii) Register or refuse registration to market functionaries and renew, suspend or cancel such registration, supervise the conduct of the market functionaries and enforce conditions of registration;

iv) Regulate or supervise the auction of notified agricultural produce in accordance with the provision and procedure laid down under the rules made under this act or bye-laws of the market committee;

v) Conduct or supervise the auction of notified agricultural produce in accordance with the procedure laid down under the rules made under this act or bye-laws of the market committee

vi) Regulate the making, carrying out and enforcement or cancellation of agreements of sales, weigh men, delivery, payment and all other matters relating to the market of notified agricultural produce in the manner prescribed;

vii) Provide for the settlement of all disputes between the seller and the buyer arising out on any kind of transaction connected with the marketing of notified agricultural produce and all matters ancillary.

viii) Take all possible steps to prevent adulteration of notified agricultural produce; make arrangements for employing by rotation, weigh men and hammals for weighing and transporting of goods in respect of transactions held in the market yard/ sub yards (maoria and joshi, 1995)

2.3: Research Gap

- Very rare research consumer opinion has been consider for the channel effectiveness that is most essential in Odisha context where the consumer prefer to purchase vegetable directly from consumer on the dimension of freshness.
- Very rare research both the consumer and marketers view point has taken for the study to measure the vegetable market efficiency.
- Very rare research has done by taking the consideration of demographic factors of consumers and farmers.
- Very rare research conducted focusing to the regulated marketing, especially vegetable market which, is addressed in the current research.

Research gap Identification

Table -2 : literature review (objectives, Methodology, findings and research gap)

SL No	Title of the article/ and Author	Objective (s)	Methodology	Major findings	Research Gap
1	A Study on "differentiator in Marketing of fresh fruits and Vegetables from Supply Chain Management Perspective". Deliya, M., Thakor, B., & Parmar, B. (2011).	To know the significant role of Supply Chain for Marketing of Fruit and Vegetables To know the role played by the various intermediaries in Supply Chain.	The secondary information was collected from the report of National Horticultural Department of India, Standard books, International Journals, articles and websites were also referred to. The present study is based on information collected from: the market officials of the selected fruits and vegetable markets, commission agents/ wholesalers, retailers and farmers in and around the selected areas	In current unorganised retail scenario, the drawback is long distribution channels, which makes unorganised retailer to sale F&V to consumer at high price. The unorganised distribution channel have more wastage due to number of time load and unload and not having scientific handling and packaging. There is imbalance between the supply and demand due to poor forecasting, which boost the price in some months or seasons. And also the vegetables are not plucked from the firm due to the less demand.	No where consumer opinion has been considering for the channel effectiveness that is most essential in Odisha context where the consumer prefer to purchase vegetable directly from consumer on the dimension of freshness Rarely any research / study is addressing the view point of middlemen regarding govt. or govt. undertaking machinery function and there principle. Very rare research has done by taking the consideration of demographic factors of consumers and farmers Very rare research conducted focusing to the regulated marketing, especially vegetable market which, is addressed in the current research.

2	Study of existing vegetable marketing channel with special emphasis on agricultural produce marketing committee (APMC), Dumbhal, Surat. Mehta, B.M. and Sonawane, Madhuri (2012).	To study the marketing practices of farmers of Surat district of Gujarat state. The study has revealed many interesting observations on the marketing of vegetables and some critical aspects relating to the marketing operations	Primary data collected from the farmers, agencies and commission agents. Data analysed through with the help of simple statistical technique using pie-charts and tables.	Most of the farmers are using commission agent to sales their products, by which around 57% of price paid by the consumer are received by the commission agents. Marketing cost of produces are mainly affected by the perishability of produces, breakage, spoilage, grading, transportation, storage, unfair and wasteful trade custom Seasonal demand and supply which multiply the cost of produces that makes variation in price paid to farmer for their produce.	
3	Organised Retailing of Vegetables: A Farmer, Retailer and Consumer Perspective in India Aparna, B., and Hannumaanthaiah, C. V.(2013).	To assess the impact of the upcoming supermarkets on farmers, consumers and different marketing channels in Rangareddy district of Andhra Pradesh which is operating their Retail outlets in Hyderabad city.	The data for the study was collected from farmers, retailers and consumers through an interview schedule in the agricultural year 2010. Tabular analysis was carried out to examine the socio economic profile of the selected farmers, farmer's productivity and profitability, marketing aspects.	The net price received by the farmer in supermarket channel was more compared to traditional channel. The total marketing cost incurred by growers was more in traditional channels than supper market channel. The producers' share in consumers' rupee is higher for supermarket channel than traditional market channel.	

4	Business Models of Vegetable Retailers In India Rajkumar, P., & Jacob, F.(2010).	To study the business models of vegetable retailers in both the organised and unorganised retailing.	Exploratory method of study is used. Data collected from 9 organised retailer, 27 wholesalers, 20 commission agents, 52 traditional retailers, 115 consumers by the personal interview questioner methods. Questioner are open ended and semi-structured. Simple statistical technique used for study.	Organised retailer usually gives higher price and make faster payment to farmer than traditional retailer. Organised retailers buying centre are closer to the firm-location which reduce the transportation charge for farmer. In organised retail loading and unloading charge also bear by retailer.	
5	Scope of Supply Chain Management in Fruits and Vegetables in India Rais, M.Sheoran A.(2013)	To find out the gap between per capital demand and supply due to enormous wastage during post-harvest storage and handling.	Primary data collected from the retailer and secondary data collected from the different government and non-government source. Simple statistical technique used for data analysis.	There is huge wastage during the post-harvest storage and handling due to improper bagging without crating, lack of temperature controlled vehicles, no cold chain facilities for preserving the produce.	
6	Indian vegetables: production trends, marketing efficiency and export competitiveness Dastagiri, M. B., Chand, R., Immanuelraj, T. K., Hanumanthaiah, C. V., Param	To estimates production trends, market efficiency and export competitiveness of vegetables in India and suggest measures to improve production,	20 crops were studied and for each crop 120 farmers were selected. The data pertains to the year 2009-10 Analytical techniques were employed are functional, logistic model, rank correlation and	Marketing efficiency is highest in 'producer to consumer channel' as compare to other. In case most of the commodity, marketing cost, marketing margin, and transport cost, labour charges are adversely affect	

	sivam, P., Sidhu, R. S., ... Ganesh Kumar, B. (2013).	marketing and exports of Indian vegetables.	Delphi technique Shepherd formula, Acharya Modified Marketing Efficiency formula were used for the estimating marketing efficiencies.	ing marketing efficiency and open market price.	
7	Issues and challenges in the supply chain of fruits & vegetables sector In india: a review Negi, S., & Ananda, N.(2015)	To identify the factors affecting supply chain of Fruits & Vegetables sector in India. To suggest mitigation strategies for the identified challenges in Supply Chain of Fruits & Vegetables sector in India	Descriptive research has been used for this study. The literature has been divided into various themes according to the issues in the supply chain and an investigation has been attempted to identify various factors affecting the supply chain.	The factors which affect the overall growth of the firm and agricultural development of India are the cold chain facility, fragmented supply chain, integration between partners, taxation issue and cost of packaging of material. Further, technology and technique, farmers' knowledge and awareness, quality and safety standards, processing value addition and market information etc. are the factors also affect the overall growth of agricultural development in India.	
8	Food exporters and Cooperation relationship: an analysis on the vegetable supply chain Gommez, E. G., Mesa, J. C. P., and Giagnocavo C. L.(2014)	To find out influence of co-coopetition on food exporting in different distribution channels, taking as reference the	The study analyses the data collected from vegetable exporters' associations and the firms' individual financial reports. A multivariate empirical model is developed to wholesalers.	Cooperation with supplier of firm's product (vegetable) can have positive effect on risk facing and improve the export propensity.	

		vegetable farming marketing Sector in southeast Spain.	Measure the impact of cooperation and competition relationships (both horizontal and vertical ones) on exporting. This model includes the influence of main buyers distinguishing basically between retailers and		
9	An Analysis of Marketed Surplus and Price Spread of Brinjal in Western Uttar Pradesh Joshi , G(2011)	To estimate the surplus of brinjal of different firm category. To determine the price spread for Brinjal in different firm category.	Cluster sampling technique is taken , three division are taken and from each division 5o vegetable grower are taken as sample Primary data are collected by the personal interview method. Simple statistical tools were used such as average percentage . Respondent are classified as marginal, small, medium and large farmer having cultivated area <1ha, 1-2 ha, 2-4 ha, >4ha respectively.	Marketed surplus of the medium firm is more as compare to the other three category of the brinjal grower. Share of producer in consumer price is high in case of the less number of the intermediaries.	

10	Rythu Bazaar: A Study of the Supply Chain of the Farmers' Markets of Andhra Pradesh Dey, S.(2012)	To study the rythu bazaar (a direct marketing place for farmer and consumer) and look at the area , if address them the firmer income generation can increase.	With analysing the secondary data, a case study is done hereby. .	In RYTHU BAZAR (A place for direct transaction between the farmer and the consumer) farmers are less affected by the middlemen and the wastage of vegetable is less as compare to commonly used marketing cannel Some improvement identified for the income generation of farmers like, facilitating the transportation for farmer, building a storage place and develop proper information system.	
11	A Comparative Study on Profitability of Supply Chain Formats in Vegetable Marketing in Karnataka Narasalagi, V.M., Hegade G.A.(2013)	To compare the profitability of supply chain viz., Traditional, cooperative and Morden supply chain format. To analyse the marketing efficiency of supply chain formats.	Data are collected from both the primary and secondary data sources. Primary data are collected from 45 farmers 4 intermediary's, 15 retailer and 60 consumers selected from supply chain model. Parentage study is done to analyse the data	About 60% of price hike without any value addition by the middle men and farmer getting 35-65 % from consumer rupees. Morden supply chain (Fab Mall, Subiksha, Trinethra, Namdhari Fresh, reliance Fresh) was found to be more effective than the cooperative and traditional supplier and again found that higher total return and net return were found in co-operative retail format than the Morden retail format	

12	Value stream analysis of vegetable Supply chain in Bangladesh: a case Study Karim, R., and Biswas, J.(2016),	To investigate the existing vegetable supply chain through the value stream analysis and identify the different stack holder and their activity. To illustrate various cost price movement towards different stages of supply chain.	The value chain study follows both quantitative and qualitative methods for data collection. the process is fully participatory through ensuring maximum involvement of farmers, wholesalers, supply chain intermediaries, retailer, and customer. However, the research paper considered a mix of the in-depth interview, focus group discussion, projective technique, and observation technique This case study is conducted in Jessore towards Dhaka for the vegetable supply chain.	The market intermediaries are categorised in to three groups 1) value added, 2) non value added, 3) necessarily value added activity. About 64% of activity is non-value activity because of redundant in nature of non-value added activity which leads to price rises up to double of producer margin. Producer and consumer are not getting proper justice due to the inadequate access of the market information and underdeveloped infrastructure.	
13	A Comparative Study of Production and Marketing Practices of Vegetables in Nainital and U.S. Nagar Districts of State Uttarakhand, India Gunwant, V. K., Hussain, M, Purohit, R. C., Ali, S. M. F., Rana, D. (2012)	To identify the existing marketing channels adopting by vegetable producers & work out marketing efficiency and marketing cost across the all channels To identify the factors affecting	Cluster sampling method and structured questioner are used for primary data collection. Secondary data collected from the district horticulture office. Simple conventional method of tabular analysis was used for study of	Lack of information regarding the market price. There is no farmer association for advocacy. Lack of policy regarding price fixation and crop insurance.	

		vegetable production and marketing.	pricing efficiency. Percentage study is done to analyse the data.		
14	Marketable Surplus and Marketing Efficiency of Vegetables in Indore District: A Micro-Level Study Pramanik, R. and Prakash, G.(2010)	To estimate the marketable and marketed surplus of vegetables grown by the Farmers in Indore District. To assess the price spread and market efficiency in marketing the principal Vegetables through different channels.	Primary data collected through personal interview method and designed schedule. Tabular analysis used for estimate marketed surplus, the quantity sold to different agencies and marketing channels.	The farmers share on consumer's rupees is low due to unreasonably high price fixed by middlemen which discourage farmers to increasing their marketable surplus.	
15	Marketing System of Agricultural Cooperatives in Nepal: A Case Studies of Janagarathi Vegetable and Fruit Producer Agriculture, Cooperative Niroja, P.,Mamoru, I., and Muto, Y.(2015)	to find out the marketing system of Agriculture cooperatives in Nepal. The specific objectives were: to identify the average sales price, marketing cost and marketing channel of vegetables sales through cooperative in Nepal.	Data collected through statured question, field observation, and personal interview were tools used. Simple statistical technique used for the data analysis.	Cooperative play important role in equal distribution among the farmers as income disparity. The share of consumer rupee is highest in case of co-operative involved marketing.	

16	The Economics of Fruit and Vegetable Marketing by Smallholder Farmers in Murehwa and Mutoko Districts in Zimbabwe Najaya, T. (2014)	To identify fruits and vegetables produced by smallholder farmers of Mutoko and Murehwa districts. To identify distribution channels of fruits and vegetables in Mutoko and Murehwa districts. To identify constraints that may prevent smallholder farmers of fruits and vegetables from marketing their products effectively.	This case study was undertaken with small-scale producers of fruits and vegetables. Qualitative data were collected through in-depth interviews, direct observations and document reviews. A structured household questionnaire was used as the basic tool to collect socio-economic, production and marketing data pertaining to fruit and vegetable smallholder producers	Farmers are facing losses due to the poor infrastructure for storage, transportation and marketing facilities for fruits and vegetables of farmers. Presence of informal middlemen is main cause of reduction of profit margin of farmers.	
17	Comparison of farm production and marketing cost and benefitAmong selected vegetable pockets in Nepal. Pokhel, D. M.(2010).	To study the proper policy formulation and program planning regarding vegetable production and marketing promotion	Snowball and purposive sampling technique used. Informal interview used for data collection. Secondary data collected from the districts agricultural development offices Simple statistical technique is used to analyse the Quantitative data by	Due to perishability nature of firms' product and lack of proper storage, firms have weaker price negotiation as compare to trader. High profit margin taken by the Trader due to the poor information of price margin as compare to the trader.	

			using MS excel. Qualitative data analysed in descriptive approach (SWOT analysis)		
18	Supply Chain Dynamics of Indian Agriculture Prawez, S(2014)	To study the current status of infrastructure in the context of information technology in Indian Agriculture. To assess the impact of inadequate agricultural Infrastructure on supply chain and agriculture.	Analysis based up on secondary data collected different source. Qualitative research analysed in descriptive approach.	ICT(information and communication technology) help the farmer how much to produce and when to produce and also help to decide the channel through which the produce flow. The major issue of distribution of vegetable is lack of integration among the channel members. Each participant in the chain act as an independent agent with low level of business relationship.	
19	Food supply chain management in Indian Agriculture:Issues, opportunities and further research Prawez S.(2014)	To assess the importance of supply chain in Indian Agriculture. To explore the issues, importance and scope of further research in Indian Agriculture.	Primary data collected in observation method and analysed in case study method. Secondary data collected from different government publish reports and article published by individual. Qualitative research analysed in descriptive approach.	To reduce post-harvest losses and deliver fresh vegetable to the end user there should need the collaboration between the supply chain partners. ICT(information and communication technology) plays an important role for farmers to decide what to produce, what are the value should have added after harvest, and which channel	

				should be choose for the flow of produces. Most of the issue are faced by farmers are lack of infrastructure like, insufficient cold chain, not good transportation facility to the firm, proper market choice and lack of information.	
20	Impact of marketing channels on perceptions of quality of life and profitability for Wisconsin's organic vegetable farmers. Silva, E., Dong, F., Mitchell, P., & Hendrickson, J. (2015).	This study aims to fill in this gap by quantitatively examining the determinants of market channel choices and the connections between market channel choices and farmer satisfaction with farm profitability and quality of life for certified organic vegetable Producers in Wisconsin. To provide more informed programmatic, educational and policy decisions	This study used a mail survey to collect information on farmer satisfaction with their quality of life and farm profitability and factors believed to be important covariates. The survey asked farmers for farmer demographic information and farm Characteristics. farmers were asked to rate their satisfaction with them farm's profitability and with their quality of life as very dissatisfied, dissatisfied, neutral, satisfied or very satisfied. Statistical tools used for	Direct marketing channel falling to achieve the goal because, insufficient in production, unwillingness of the customer to pay the price needed by the farmer to cover their cost More broadly across market channels, it appears that perceptions of profitability influence a farmer's perception of their quality of life. As such, efforts to assist farmers to increase their profitability in market channels that also provide a higher quality of life would be beneficial. More education and training related to farm finances, thus allowing for more accurate assessment of the profitability of selling	

		regarding the support of direct market channels.	study are factor analysis and regression.	into different market channels as it relates to their specific perceptions of quality of life and profitability.	
21	A study on problems of marketing vegetables In farmers market. Kumar , m.p.(2012).	To analyse of problems of vegetables marketing in Tamilnadu on the basis of ratings of the vegetables growers.	Stratified random sampling method is used for data collection, where three strata are taken. Data are collected with the with help of well-structured interview schedule. The data interpretation is done with the help of average and ANOVA analysis.	It is concluded that the respondents rate high level problems of marketing vegetables and fruits in farmers' market with reference to damage cost, intermediaries' exploitative practices, perishability of product, transportation cost and high storage cost. The respondents rate moderate level problems of marketing vegetables and fruits in farmers' market with reference to freight charges, lack of proper grading, high carriage and other handling charges, exploitation of growers by market force, lack of proper quality control, long distance of market access, seasonal of production, long marketing channel, delay payment, lack of cold storage place, advance sales agreement, inadequate post-harvest care, monopoly of middleman, bulkiness of products and low exports. The respondents rate low	

				level problems of marketing vegetables and fruits in farmers' market with reference to irregular supply, primitive method of selling and price fixation, packing and loading problems, quality variation in production, packing of products. There is an inverse relation between the farm size of the respondents and their overall problems of marketing vegetables and fruits in farmers' market	
22	An overview of the supply chain management of malaysian vegetable and fruit industries focussing on the channel of distribution. Man, N., Nawi, M. N., Islam, M. M.(2009)	To analyse the role of channel of distribution of the fruits and vegetable malaysian market.	The primary data for this study were collected through a market survey using face-to-face interview with selected market participants such as retailers, wholesalers, transporters, processors, packers, assemblers and the producers stratified random sampling used for the data collection. Questioner method used for the data collection. To analyze data measures such as the mean, standard deviation, frequency distribution and variability are calculated	More than half of the fresh fruits and vegetables produced flows through the wholesalers. Movement and distribution of fresh fruits and vegetables from the farm to the retailers is dominated by wholesalers. Consumer preferences as well as the consumption patterns and styles are also changing due to easy access to information. Markets are adjusting themselves by responding to consumers' needs and wants.	

23	The fruit and vegetable Marketing chains in Thailand: Policy impacts and implications Srimanee, Y., Routray, J. K(2012)	The main objective of this paper is to identify and study the marketing chains of fresh fruit and vegetables (FFV) in Thailand, the linkages of FFV farmers with supermarkets, and an evaluation of supermarket procurement systems. To discuss the policies and implications relating to FFV and the connection of farmers with supermarkets in the context of Thailand.	The paper uses both secondary and primary data. Public agencies' documents and published reports were used to describe retailing and FFV policy of Thailand. Primary data were collected using a field survey, observations in different markets, interviewing key informants (farmers, intermediaries, wholesalers, supermarket managers, cooperative representatives and government officers) using a checklist and farmer survey in 2010. Two sided chi-square test and the Fisher's exact test was used to analyse the data	The government has made various attempts to improve the FFV market by promoting both domestic and export markets simultaneously thus improving production efficiency. Supermarkets play important roles in connecting farmers to markets through direct procurement while improving cultivation practices of participating farmers in order to enhance quality of produce. A number of policy gaps still exist to protect the interests of the farmers in maximizing their benefits, and in the areas of participation between public and private sectors.	
24	Freshness of fruits and vegetables: consumer language and perception Pe´neau, S., Linke, A., Escher F. & Nuessli, J.(2009)	To identify descriptions involved in defining the concept of freshness from a consumer perspective.	A total of 559 subjects recruited in supermarkets of different cities and villages in Switzerland were asked to write down their understanding of freshness in general, and for fruits and vegetables in particular.	Respondents mentioned a large variety of descriptors in relationship with fresh fruits and vegetables. Results suggest that freshness describes a level of closeness to the original product, in terms of distance, time and processing.	

			Direct elicitation by means of an open-ended questionnaire was used. Two sided chi-square test and the Fisher's exact test was used to analyse the data.	Respondents having more contact with the place of fruit and vegetable production mentioned non-sensory attributes more frequently, whereas those having fewer contacts mentioned sensory attributes more frequently. It can be concluded that consumers have a collective concept of freshness but they use their different everyday experiences with fruits and vegetables to describe freshness.	
25	Agri-fresh produce supply chain management: a state-of-the-art literature review Shukla , M. & Jharkharia, S(2013)	To present a literature review of the fresh produce supply chain management (FSCM). FSCM includes the processes from the production to consumption of fresh produce (fruits, flowers and vegetables).	Literature review is done by systematically collecting the existing literature over a period of 20 years (1989-2009) and classifying it on the basis of structural attributes such as problem context, methodology and the product under consideration. The literature is also categorized according to the geographic region and year of publication.	There is an increase in interest towards FSCM still there is an absence of a journal with the prime attention towards FSCM. The key finding of this review is that the main interest is towards consumer satisfaction and revenue maximization with post-harvest waste reduction being a secondary objective. It is revealed from the review that most of the literature is fragmented and is in silos. Lack of demand forecasting, demand and supply mismatch, lesser integrated approach etc are the major causes of concerns.	

26	The fruit and vegetable marketing chains in Thailand: policy impacts and implications Srimanee, Y & Routray, J. K.(2012)	To identify and study the marketing chains of fresh fruit and vegetables (FFV) in Thailand, the linkages of FFV farmers with supermarkets, and an evaluation of supermarket procurement systems. To discuss the policies and implications relating to FFV and the connection of farmers with supermarkets in the context of Thailand.	The paper uses both secondary and primary data. Public agencies' documents and published reports were used to describe retailing and FFV policy of Thailand. The supermarket managers and chain stakeholders were interviewed about the FFV marketing chains and supermarkets' procurement system. Policy issues were discussed with policymakers, and extracted from relevant documents.	The government has made various attempts to improve the FFV market by promoting both domestic and export markets simultaneously thus improving production efficiency. Supermarkets play important roles in connecting farmers to markets through direct procurement while improving cultivation practices of participating farmers in order to enhance quality of produce. A number of policy gaps still exist to protect the interests of the farmers in maximizing their benefits, and in the areas of participation between public and private sectors.	
27	Vegetables procurement by Asian supermarkets: a transaction cost approach Ruben, R., Boselie, D. & Lu, H.(2017)	The paper seeks to understand the conditions that motivate Asian supermarkets' choices for vegetable sourcing through wholesale procurement or preferred supplier systems.	Insights from transaction cost theory are used to analyse the evolution of fresh produce procurement regimes by Supermarkets in Bangkok and Nanjing.	Trade-offs between higher production cum-investment costs (i.e. fixed investments, variable production costs and economies of scale) with expected savings in transaction costs (governance and opportunistic behaviour) are registered that could hinder contractual delivery.	

28	Food exporters and co–operation relationships: an analysis on the vegetable supply chain. Gómez, E. G., Mesa, J. C. P. & C. L.(2015)	To provide empirical evidence on the influence of co-opetition on food exporting in different distribution channels, taking as reference the vegetable farming marketing sector in southeast Spain.	The study analyses the data collected from vegetable exporters' associations and the firms' individual financial reports. A multivariate empirical model is developed to measure the impact of cooperation and competition relationships (both horizontal and vertical ones) on exporting. This model includes the influence of main buyers distinguishing basically between retailers and wholesalers.	The results obtained show that diverse forms of collaboration with competitors, in both horizontal (such as logistics and research projects) and vertical dimensions (such as promotion and quality certifications) have positive effects on vegetable export propensity. These influences become more apparent when large retailers are the main buyers.	
29	A multi-stakeholder perspective on the adoption of good agricultural practices in the Thai fresh produce industry Wongprawmas, R., Canavari, M.,& Waisarayutt, C.(2015)	To explore the factors hindering the adoption of good agricultural practices (GAPs) in the Thai fresh fruit and vegetable industry from the perspectives of key stakeholders in different tiers of the supply chain up to consumers.	Semi-structured interviews were carried out in Bangkok and nearby cities. The key informants were experts, governmental authorities, producers, distributors, exporters, and consumers. A qualitative content-summarising approach was used to analyse the interview transcripts.	Stakeholders perceived the credibility of national GAP (Q-GAP) standards implemented in the Thai domestic market as low, due to a lack of reliable verification and traceability and because the system is managed by governmental bodies. Consumer demand could drive the market of safe produce, but retail stores are the main influencers of producers' adoption of GAPs in domestic markets	

30	Modern food retailing buyingbehaviour in Africa: the case of Tanzania Nandonde, F. A. & Kuada, John(2016)	To explore modern food retail buyers' behaviour in developing economies using the case of Tanzania.	A qualitative case study approach was used for the study. The researcher used semi-structured interviews with retailers for data collection and corroborated them with secondary data. Data were thematically analysed.	The study shows that the criteria used by modern food retailers in the selection of local food suppliers are reliability, quality, trade credit and legal certification. The task is further complicated by the overlapping food certification requirements of various government agencies, which impose limitations on the buyers' decision. Due to the exploratory nature of the study and its focus on the context of a particular geographical marketplace, the findings may not be generalising to other countries.	
31	Comparison of relationship perception in the internationalfresh produce business Gyau, A. and Spiller, A(2009)	to compare the perceived relationship quality of the retailer group of buyers with the buyers in the other categories of the fresh fruit and vegetable business between Ghana and Europe.	Multivariate analysis of variance (MANOVA) procedure and post hoc test were used to determine the differences in the perceived relationship quality of the exporters classified according to their main European buyer type	In the context of the fresh produce business, the level of perceived power exploitation may differ depending on the type of business organisation or the position of the firm in the supply chain.	

CHAPTER - 3

Research Design and Methodology

As the research design is the blueprint of the entire research, the type of research design, sampling, data collection instrument, scale used in the instrument, process of executing the instrument, the respondents profile, the data analysis tools, techniques and procedures are defined as follows.

3.1 Research Design:

The current study will follow causative research design, where the relationships of satisfaction of farmers-cum-businessmen and the satisfaction of customers with the factors of marketing efficiency simultaneously are examined. This relationships justified by a defined objectives and hypotheses. From existing literatures, the research gap is developed. The research gaps lead to the objectives and hypotheses of the study.

3.2 Sampling:

The data is collected through stratified random sampling method. Here the stratification is made on the basis of 3 major geographical areas of Odisha. So, 3 strata as 3 major town areas of Odisha like Sambalpur & Bargarh , Bhubaneswar & cuttack and Angul & Dhenkanal. Two set of questionnaires are developed for two different samples of respondents like farmer-cum-businessmen (sample-1) those who sale in RMC market and the vegetable consumers (sample-2) those who are purchasing from the RMC market. From each stratum for the sample-1, 50 respondents are executed with the structured questionnaire. From each stratum for the sample -2, 50 respondents are executed with the structured questionnaire in online platform. So, the sample size for each sample is 150. The total sample size is 300 spreading over two set of questionnaires.

3.3 Respondents' Profile

Two groups of the respondents are taken for the data collection are farmer-cum-businessmen and consumer. Where, 150 respondents are taken from farmers-cum-businessmen and 150 respondents are taken for consumers.

Out of 150 respondent of farmers-cum-businessmen 90% men and 10% female, 20% single and 80% married. As qualification concern 63.3% of respondents are below matriculation,26.0% of respondents are +2 level and 10.7% of respondent are +3 level. Whereas the 24% of the farmer-cum-businessmen are involve in their business himself only and 39.3% of the farmers-cum-businessmen's 2 family member involve in their profession, 25.3% respondents have 3 family members are involving in their profession and 11.3% respondent have 4 or more family members are involving in their business.

Whereas, out of 150 respondent of consumer 3.3% respondents are below 20 years, 40.7% respondents are between age 21-30, 41.3% respondents are between age 31-40, 6.7% respondents are between the age 41-50 and 8% respondents are age above 50 years. 68 % of respondents are male where as 32% of the respondents are female consumers. As marital status concerns 63.3 % married consumer and 36.7% respondents are unmarried consumers. And, only 4% consumers are below matriculation, 16% respondents are +2 level, 37.3% respondents are +3 level and 42.7% respondents are P.G level as educational qualification concern.

3.4 Data collection:

Two set of structured questionnaire are prepared keeping in view the defined objectives and hypotheses. Questionnaires followed five point Likert scale and executed to the above said samples. Questionnaire meant for farmers-cum-businessmen is executed in sample-1 with a face to face interaction by the researcher himself and another questionnaire, which is meant for the vegetable consumers is executed through online survey instrument. Total numbers of items in the questionnaire-1 that is prepared for farmer-cum-businessmen are 44 including 5 demographic items and total numbers of items in the questionnaire-2 that is prepared for consumers are 49 including 6 demographic factors.

3.5: Data analysis procedure& techniques

Data is analysed by various descriptive statistics and empirical methods. First, the survey instrument, questionnaires are tested for their scale reliability by Cronbach's alpha test. Two separate explorative factor analyses are adopted to extract the factors of farmers' satisfaction and consumers' satisfaction separately. Multiple linear regression method is followed to examine the degree of the farmer's satisfaction influencing the farmer's happiness and the degree of influence of consumers' satisfaction on vegetable consumers' positive orientation in the RMC markets. Famers' happiness and consumers' positive orientation integrally considered as 'marketing efficiency' in the present research. Three-way ANOVA technique is used to differentiate the farmers-cum-businessmen's happiness over the demographic factors of farmers-cum-businessmen and to differentiate farmers' happiness over the marketing habits of farmers-cum-businessmen's. The MANOVA technique is used to judge the interactions of demographic variables with the consumers' positive orientation (two dependent variables) in the RMC market and the interaction of purchasing habits with the consumers' positive orientations.

CHAPTER - 4

Data Analysis And Interpretation

4.1 Descriptive Statics

The normality of data is justified by various descriptive statics such as variance, Std. Deviation, Skewness and kurtosis and the results are shown in the bellow table-3 and table - 4.

Table -3 Descriptive statics (sample-1)

Sl. No	Variables	mean	variance	Std. Deviation	Skewness	Kurtosis
1	In a month coming to market	3.09	1.462	1.209	.178	-.430
2	Distance of the residence from the market	2.32	1.870	1.368	.919	-.035
3	Vegetable is good for health	4.03	1.106	1.052	-.979	.194
4	Vegetable are main source of income	3.86	1.383	1.176	-.477	-1.305
5	Distance of firm location and market is manageable	3.79	1.243	1.115	-.688	-.466
6	Fell entrepreneur in profession	3.89	1.269	1.126	-.516	-1.051
7	Timely enable sale vegetable	3.56	.798	.894	-.154	-.171
8	Chose RMC to sale as first preferences	3.93	.887	.942	-.683	-.113
9	Less profit margin due to more marketing cost	3.63	.799	.894	-.388	-.037
10	Getting information about market	3.13	1.199	1.095	.025	-.934
11	Can see price increase or decrease in near future	3.44	.933	.966	-.349	-.472
12	Visualizing the customer demand for product in this market	3.55	.853	.923	-.288	-.532

13	Present production meets the customer demand	3.81	.931	.965	-.570	-.366
14	Proper storing facility of pick season of the vegetable production	2.38	1.861	1.364	.569	-1.036
15	Getting adequate infrastructure in RMC market	3.20	1.289	1.135	.212	-.860
16	Fell customer are satisfied with the behavior	4.25	.620	.787	-.901	.417
17	Enable maintain good relationship with consumer	4.20	.805	.897	-1.310	2.027
18	Adding value before selling to attract consumer	3.47	1.244	1.115	-.667	-.227
19	Facing problem of loading and unloading	2.99	1.530	1.237	.090	-.909
20	Having proper planning for selling product in RMC market	3.56	1.040	1.020	-.415	-.551
21	Prefer sale product as whole at a time	3.22	1.515	1.231	.009	-1.025
22	Think little-bit risk of selling vegetable in the vegetable	4.08	.987	.993	-.953	.256
23	Suffer marketing surplus in this market	3.01	1.194	1.093	-.027	-.661
24	Getting proper place of selling of vegetable in RMC market	3.27	1.838	1.356	-.247	-1.132
25	Commission agents are affect negatively in the business	3.23	1.062	1.031	.054	-.752
26	Getting expected price at the high production seasons	2.75	1.301	1.141	.168	-.685
27	For transportation of vegetable Government provide temperature control vehicle	1.21	.317	.563	2.763	7.190
28	Prefer to sale unsold product to commission agents	3.29	1.656	1.287	-.417	-.894

29	Wastage of product reduces the profit margin	3.51	.735	.857	-.496	.041
30	Consumer have sufficient knowledge about Vegetable	3.76	.814	.902	-.727	.353
31	Easily avail private transportation	3.91	1.427	1.195	-1.014	.107
32	Grading of product	3.77	.972	.986	-.623	.129
33	Perishability of vegetable detracting farmers to produce vegetable for selling propose	3.56	.973	.986	-.467	-.020
34	Getting reasonable profit from vegetable	3.44	.946	.973	-.361	-.324
35	Happy with the profession	3.75	.831	.912	-.351	-.628
36	Continue to sale in RMC market	4.03	1.039	1.019	-.722	-.493
37	Age of respondent	5.23	3.831	1.957	.745	.336
38	Gender	1.10	.091	.301	2.694	5.327
39	Marital Status	1.80	.161	.401	-1.515	.300
40	Average amount of sales	1.49	.789	.888	2.293	5.565

Descriptive statics such as mean, variance, standard deviation and kurtosis are described for the sample -2.

Table 4 Descriptive statics (sample-2)

Sl. No	Variables	Mean	variance	Std. Deviation	Skewness	Kurtosis
1	How many time in a month coming to this market to purchase vegetables	2.40	2.013	1.419	1.063	.437
2	Distance of the residence from the market	2.09	1.543	1.242	1.303	1.405
3	The product(s), you are generally purchasing from RMC	2.04	.602	.776	-.070	-1.330
4	Now a day vegetables are essential for good health	4.27	.804	.897	-.795	-.720
5	I enjoy purchasing from the RMC market	3.83	1.308	1.144	-.759	-.188
6	Getting more variety of vegetables from the RMC market	4.10	1.044	1.022	-.739	-.585

7	RMC market is convenient to purchase the vegetables.	3.81	1.133	1.064	-.567	-.232
8	Getting vegetable at reasonable price from the RMC market	3.53	1.285	1.133	-.221	-.764
9	Like to purchase the vegetable from the farmer directly rather than vegetable merchant	3.82	1.209	1.100	-.434	-.802
10	Getting fresh and qualitative vegetables at RMC market.	3.37	1.189	1.090	-.064	-.781
11	Can see the price-increase or decrease of vegetables for the near future	3.65	1.114	1.055	-.099	-1.088
12	Farmers-cum business men are behaving properly in RMC market with the general customers	3.24	1.177	1.085	-.012	-.433
13	Purchasing getting more verity of product from one farmer	3.23	1.455	1.206	-.088	-.821
14	Getting appropriate information regarding vegetable from this market.	3.24	1.231	1.109	-.010	-.578
15	Think the farmers-cum sellers are treating equally to all general customers in this RMC market	3.25	1.355	1.164	.023	-.825
16	Seasonal products are very cheaper at the pick production season	3.73	1.378	1.174	-.525	-.698
17	Some times the price suddenly increases as farmers-cum sellers sell the vegetables to other whole-sellers and retailers	3.79	1.158	1.076	-.331	-1.060
18	Govt. is not taking appropriate steps to regulate properly the RMC markets	3.71	1.266	1.125	-.187	-1.379
19	Organised retailer are more systematic than the farmers selling their product at RMC market	3.39	1.260	1.123	-.133	-.824

20	Know about the MSP (minimum supporting price) fixed by the Govt. and accordingly I purchase the products	2.77	1.254	1.120	.241	-.555
21	Getting vegetable at credit from the RMC market	2.49	1.701	1.304	.419	-.915
22	Use to purchase the product from same seller in most of the time	3.35	1.237	1.112	-.054	-.896
23	Use to check the quality of vegetable from different seller before purchase the product	3.63	1.334	1.155	-.357	-.765
24	Getting trade discount when I purchase bulk amount of products	3.37	1.363	1.167	.003	-1.068
25	Getting vegetables in hygienic condition in the RMC market yard	3.09	1.374	1.172	.121	-.867
26	Same quality vegetables are available through-out the day in this RMC market	3.05	1.213	1.101	.212	-.539
27	Special carry bag for vegetable are provided by the seller of vegetables in RMC market	2.73	1.472	1.213	.275	-.711
28	Generally bargain for price with the seller while purchasing the vegetable products in this market	3.38	1.365	1.168	-.192	-.697
29	Having more choices of sellers in this RMC market	3.68	1.132	1.064	-.007	-1.331
30	Seeing almost the same sellers every day in this market	3.70	1.218	1.104	-.382	-.834
31	Getting directly the firms products from the farmers-cum-sellers which can be store for several days	3.27	1.512	1.230	-.150	-.853
32	RMC market yard have good parking place	2.77	1.361	1.167	.286	-.608
33	RMC market yard is well structured for consumer facility	2.89	1.438	1.199	.090	-.686

34	Weight measurement in the RMC market seller is quite good	3.18	1.276	1.130	.063	-.623
35	Takes little more time to purchase vegetables from this market	3.36	1.024	1.012	-.064	-.389
36	Getting my product with in my budget prediction	3.37	1.162	1.078	-.074	-.603
37	I face problem to find out the organic vegetable in RMC market	3.55	1.270	1.127	-.118	-1.052
38	I continue to purchase vegetable from the regulated market	3.42	1.131	1.064	-.195	-.255
39	This market has more space for shopping activities	3.07	1.276	1.130	.081	-.643
40	Feel more rush in this market in holidays in comparison to other days.	3.74	1.268	1.126	-.301	-1.095
41	Market place is well structured for the shopping & marketing activities	3.07	1.397	1.182	.054	-.719
42	Think farmers-cum-sellers are doing their business happily in this RMC market	3.33	1.405	1.185	-.208	-.695
43	Age	2.75	.872	.934	.864	.590
44	Gender	1.32	.219	.468	.780	-1.411
45	Marital Status	1.37	.234	.484	.559	-1.711
46	Qualification	3.19	.717	.847	-.771	-.165
47	Income per month	3.10	2.010	1.418	-.165	-1.217
48	Profession	3.20	3.638	1.907	.574	-.851
49	Average amount purchase	3.85	2.936	1.713	-.018	-1.399

The above descriptive statistics refers that the two set of data(sample -1 and sample- 2) are normally distributed. So, normality assumption for the parametric tests is satisfied here. The detail statistics along the histograms are there in the annexture-3(table no-27.1 to 28.49 and Figure No- 6.1 to 7.49).

4.2 Empirical Analysis

The scale reliability of two set of data (farmers-cum-businessmen's and consumers) are tested by the Cronbach's alpha. Two separate explorative factor analysis is applied to explore the factors of these two set of data. These explored factors are taken as independent variable for the dependent variable in regression analysis, which is done separately for separate set of data. To differentiated farmers' happiness over demographic variable 3-way ANOVA has applied. Again, 3-way ANOVA also applied to differentiate farmers-cum-businessmen's happiness over marketing habits of farmers. For determining the interaction of demographic variable with the consumers' positive orientation the MANOVA applied and also MANOVA has applied to analyse the interaction of the purchasing habits with the consumers' positive orientation.

4.2.1: Scale reliability farmers-cum-businessmen's data set.

To measure the scale reliability of questionnaire, Cronbach's alpha is used. It is the measurement of the internal consistency, which shows the close relativeness of a set of items as a group.

Table 5 Scale reliability (sample -1)

Mean	Variance	Std. Deviation	N of Items	Cronbach's Alpha	Cronbach's Alpha Based on Standardized Items	N of Items
125.37	110.717	10.522	37	0.642	0.670	37

The scale (questionnaire) is optimally reliable, i,e. 67% as the standardised value of Cronbach's alpha is 0.670 with 37 items excluding 3 demographic variables, referred in the table –5.

4.2.2 Exploration of major factors of farmer-cum-businessmen's satisfaction

To identify major factors of farmers-cum-businessmen's satisfaction on vegetable production and marketing in the RMC market, the explorative factor analysis is adopted by taking 33 variables.

Table 6 KMO and Bartlett's Test : Sample Adequacy (Sample -1)

Kaiser-Meyer-Olkin Measure of Sampling Adequacy.		**.531**
Bartlett's Test of Sphericity	Approx. Chi-Square	1612.105
	Df	528
	Sig.	.000

KMO measure of sample adequacy is 0.531, which is significant (0.000). So explorative factor analysis is applicable here as referred in the table- 6.The significant (0.000) statistics(0.531) of Bartlett's test of Sphericity Justify that the correlation matrix is not an identity matrix. This shows, that the sample is very appropriate for factor analysis.

Table - 7 Total Variance Explained(Sample-1)

Component	Initial Eigenvalues			Extraction Sums of Squared Loadings			Rotation Sums of Squared Loadings		
	Total	% of Variance	Cumulative %	Total	% of Variance	Cumulative %	Total	% of Variance	Cumulative %
1	3.757	11.384	11.384	3.757	11.384	11.384	2.628	7.963	7.963
2	3.510	10.636	22.021	3.510	10.636	22.021	2.347	7.113	15.076
3	2.436	7.383	29.404	2.436	7.383	29.404	2.170	6.577	21.653
4	1.925	5.832	35.236	1.925	5.832	35.236	2.135	6.470	28.123
5	1.688	5.116	40.352	1.688	5.116	40.352	1.936	5.866	33.989
6	1.601	4.851	45.202	1.601	4.851	45.202	1.839	5.574	39.563
7	1.517	4.597	49.800	1.517	4.597	49.800	1.804	5.467	45.031
8	1.485	4.500	54.300	1.485	4.500	54.300	1.740	5.273	50.304
9	1.310	3.971	58.270	1.310	3.971	58.270	1.699	5.148	55.452
10	1.260	3.817	62.088	1.260	3.817	62.088	1.488	4.508	59.960
11	1.175	3.562	65.650	1.175	3.562	65.650	1.453	4.402	64.362
12	1.023	3.099	68.749	1.023	3.099	68.749	1.448	4.386	**68.749**
13	.990	3.001	71.750						
14	.892	2.704	74.454						
15	.845	2.560	77.014						
16	.803	2.434	79.448						

17	.778	2.356	81.804	
18	.704	2.132	83.936	
19	.658	1.994	85.931	
20	.592	1.794	87.724	
21	.554	1.678	89.402	
22	.493	1.493	90.895	
23	.446	1.352	92.247	
24	.378	1.144	93.391	
25	.344	1.041	94.433	
26	.331	1.002	95.435	
27	.322	.975	96.410	
28	.279	.846	97.256	
29	.238	.722	97.978	
30	.205	.620	98.598	
31	.176	.534	99.133	
32	.151	.457	99.590	
33	.135	.410	100.000	
Extraction Method: Principal Component Analysis.				

By principal component method and varimax rotation with 24 iterations 12 factors are extracted. The numbers of variable more than 1 initial eigenvalues are conceder as factors. These 12 factors explain near about 69% variance as cumulative percentage of rotated square loading is 68.749, referred in the table -7.

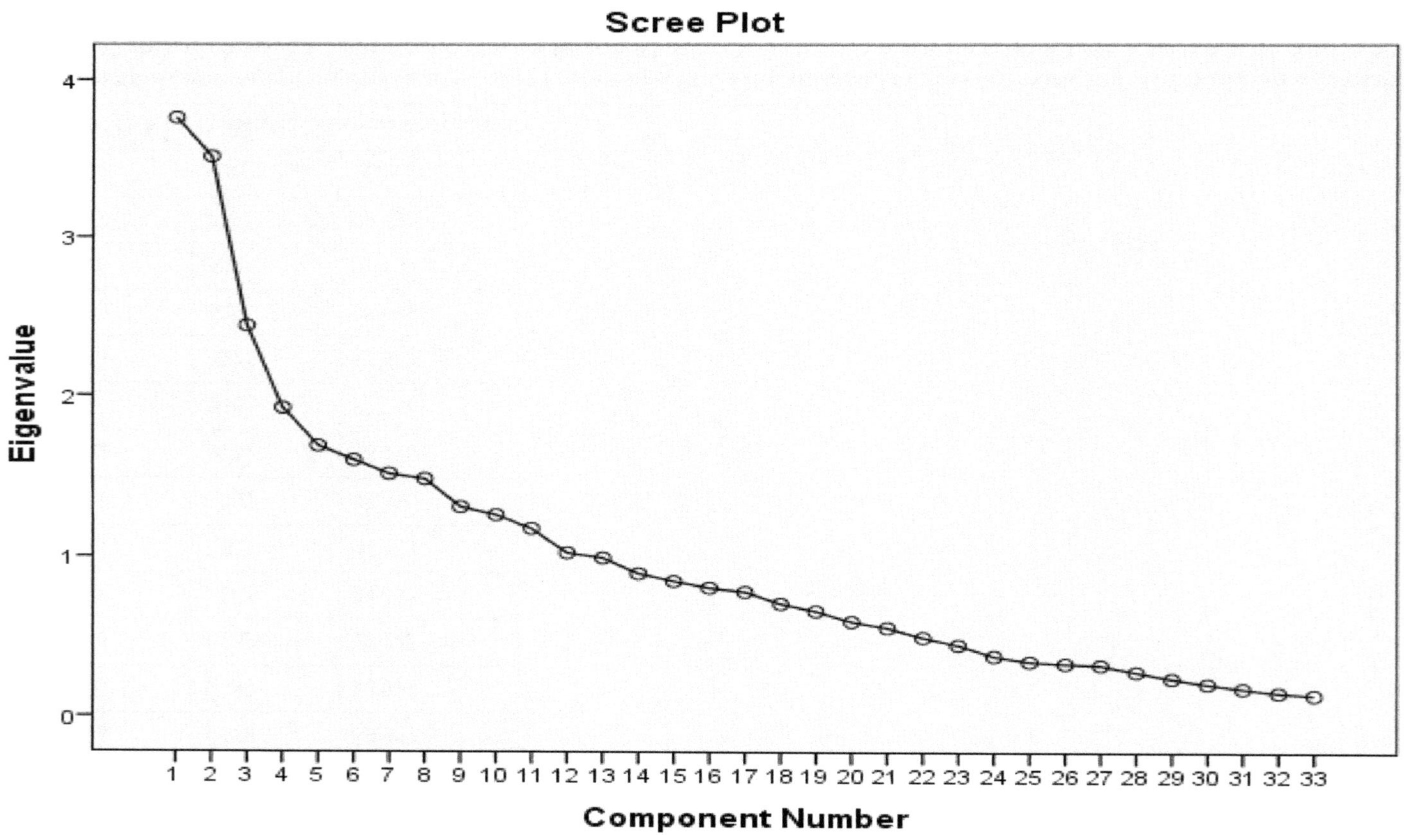

Fig. 2: Scree plot for farmers-cum-businessmen's factors

By visualising the scree plot for farmers-cum-businessmen's factors there is sudden elbow mark at the 12 component number in the figure-2 which, explain the explored factor of the factor analysis.

Table – 8 Rotated Component Matrix[a](Sample-1)

	Component											
	1	2	3	4	5	6	7	8	9	10	11	12
Getting reasonable profit from vegetable	.797											
Distance of firm location and market is manageable	.730	Farmers contentment in market										
continue to sale in RMC market	.640											
Enable maintain good relationship with consumer		.844										
Feel customer are satisfied with the behaviour		.813	Consumer relation									
For transportation of vegetable government provide temperature control vehicle		-.478										
Can see price increase or decrees in near future			.766									
visualising the customer demand for product in this market			.682	Market anticipation								
Getting adequateinfrastructure in RMC market			.600									

Statement	Factor	1	2	3	4
Adding value before selling to attract consumer	Market planning	.729			
Having proper planning for selling product in RMC market		.672			
wastage of product reduces the profit margin		-.481			
Chose RMC to sale as first preferences		-.406			
Vegetable are main source of income	Farming professionalism		.755		
Fell entrepreneur in profession			.599		
Facing problem of loading and unloading			.496		
Proper storing facility of pick season of the vegetable production	Easy of vegetable marketing			-.688	
Vegetable are good for health				.642	
Getting proper place of selling of vegetable in RMC market				-.453	
Commission agents are affect negatively in the business	Marketing cost				.764
Suffer marketing surplus in this market					.621
Less profit margin due to more marketing cost					.498

Timely abele sale vegetable	Proper value of product	.814				
Getting expected price at the high production seasons		.633				
Prefer to sale unsold product to commission agents	Convenience selling of vegetable		.758			
Easily avail private transportation			.550			
Present production meet the customer demand			.386			
Grading of product	Grading of product			.791		
Getting information about market	Product information				.789	
Prefer sale product as whole at a time					-.463	
Perishability of vegetable detracting farmers to produce vegetable for selling propose					-.443	
Think little-bit risk of selling vegetable in the vegetable	Consumer knowledge					.774
Consumer have sufficient knowledge about Vegetable						.519

Extraction Method: Principal Component Analysis.
Rotation Method: Varimax with Kaiser Normalization.[a], a. Rotation converged in 24 iterations.

By using Principal component analysis and by varimax rotation with Kaiser Normalization, 12 factors are extracted as described in the rotated component matrix table (Table-8) which are named as Farmers contentment in market, Consumer relationship, Market anticipation, Market planning in RMC market, Farming professionalism, Ease of vegetable marketing, Marketing cost, Proper value of product, Convenience selling of vegetable, Grading of product, Product information and Consumer knowledge.

4.2.3 Influence of factors of farmers-cum-businessmen satisfaction on farmer-cum-businessmen's happy in their profession.

The influence of farmers' satisfaction (12 factors) on farmers-cum-businessmen's happiness in their profession is tested by the following hypothesis and the hypothesis is tested by multiple linier regression.

H_1: There is a significance impact of factors of farmers' satisfaction on the farmers-cum-businessmen's happiness in their profession.

Table 9 ANOVA[a]: Model Fitting for H_1 (Sapmple-1)

Model		Sum of Squares	Df	Mean Square	F	Sig.
1	Regression	64.466	12	5.372	12.389	**.000[b]**
	Residual	59.408	137	.434		
	Total	123.873	149			

a. Dependent Variable: Happy with the profession

b. Predictors: (Constant), Farmers contentment in market, Consumer relationship, Market anticipation , Market planning in RMC market, Farming professionalism, Ease of vegetable marketing, Marketing cost, Proper value of product, Convenience selling of vegetable, Grading of product, Product information and Consumer knowledge

The model, multiple linear regression, is significantly fitting for this analysis as the F statistics (12.389) is significant (p=0.000), referred from the above ANOVA table (table no for 9).

Table 10 Model Summary for H_1 (Sample-1)

Mod-el	R	R Square	Adjust-ed R Square	Std. Error of the Estimate	Change Statistics					Durbin-Watson
					R Square Change	F Change	df1	df2	Sig. F Change	
1	.721[a]	.520	**.478**	.659	.520	12.389	12	137	**.000**	1.915

a. Predictors: (Constant), Farmers contentment in market, Consumer relationship, Market anticipation , Market planning in RMC market, Farming professionalism, Ease of vegetable marketing, Marketing cost, Proper value of product, Convenience selling of vegetable, Grading of product, Product information and Consumer knowledge

b. Dependent Variable: Happy with the profession

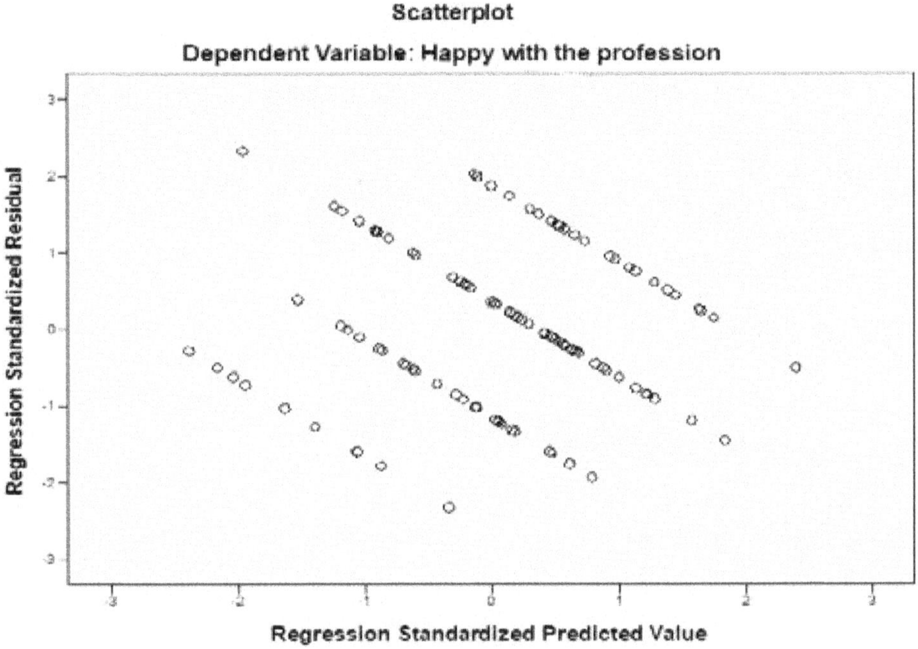

Fig. 3 : Heterosced asticity (sample-1)

The farmers-cum-marketer satisfaction factors influence farmers-cum-businessmen are happy with their profession by 48% as the significant (*P*=0.000) adjusted $R^2$is 0.478. The variables are independent because the variables have no auto co-relation symptoms avelable. It is justifying from the Durbin Watson statics (it should very from 1.5 -2.5). In current research it is 1.915 which is justify by the above argument. Referred in the table -10.

As referred in the Figure- 3 the assumption of Homoscedasticity is not violated so, the is the heteroscedasticity symptoms are not supported here. As the size of the error term differs across values of independent variables hence, the regression model is fitting here.

Table – 11 Coefficients[a]:multi-collinearity for H_1(Sample -1)

	Model	Un-standardized Coefficients		Standardized Coefficients	T	Sig.	Correlations			Collinearity Statistics	
		B	Std. Error	Beta			Zero-order	Partial	Part	Tolerance	VIF
1	(Constant)	3.753	.054		69.807	.000					
	Farmers contentment in market	.625	.054	**.686**	11.592	**.000**	.686	.704	.686	1.000	1.000
	Consumer relationship	.035	.054	.039	.654	.514	.039	.056	.039	1.000	1.000
	Market anticipation	.061	.054	.067	1.136	.258	.067	.097	.067	1.000	1.000
	Market planning in RMC market	.028	.054	.030	.514	.608	.030	.044	.030	1.000	1.000
	Farming professionalism	.113	.054	**.124**	2.092	**.038**	.124	.176	.124	1.000	1.000
	Ease of vegetable marketing	.141	.054	**.155**	2.612	**.010**	.155	.218	.155	1.000	1.000
	Marketing cost	.018	.054	.020	.332	.741	.020	.028	.020	1.000	1.000
	Proper value of product	-.017	.054	-.018	-.309	.758	-.018	-.026	-.018	1.000	1.000
	Convenience selling of vegetable	-.033	.054	-.036	-.614	.540	-.036	-.052	-.036	1.000	1.000
	Grading of product	.038	.054	.042	.711	.478	.042	.061	.042	1.000	1.000
	Product information	.006	.054	.006	.109	.914	.006	.009	.006	1.000	1.000
	Consumer knowledge	-.004	.054	-.005	-.083	.934	-.005	-.007	-.005	1.000	1.000

a. Dependent Variable: Happy with the profession

Farmers-cum-businessmen's contentment in vegetable market is an important factor as this factor contribute 69% to (β=0.686 with p=0.000) to the happiness of farmer-cum-businessmen in their profession. Another two factor Farming professionalism and Ease of vegetable marketing are also contributing 12% and 15% (β-0.124 with p=0.038 and β-0.287 with p=0.010) respectively to farmer's happiness in the farming profession. In collinearity statics shows that no multi-collinearity symptoms at all (Table-11). Of course the factors are extracted from the explorative factor analysis. So above said hypothesisis consider as supported. The value impact is moderate, so some extent it can be taken for the managerial implication.

The resulting regression model can be –

Happy with the profession= 3.753+0.686 (Farmers contentment in market) +0.124 (Farming professionalism)+0.155(Ease of vegetable marketing) + error

4.2.4 Differentiation of happy with the profession over demographic factors.

The differentiation of farmers-cum-businessmen's are happy with their profession over demographic factors (age, gender and Marital status) is tested by the bellow hypothesis and the hypothesis is tested by 3-way ANOVA.

H_2: There is significant difference of happiness of farmers-cum-businessmen's over their demographic factors.

Table 12 Levene's Test of Equality of Error Variances: forH_2(Sample – 1)

Dependent Variable: Happy with the profession			
F	df1	df2	Sig.
2.185	18	131	.006

Tests the null hypothesis that the error variance of the dependent variable is equal across groups.

a. Design: Intercept + Age + Gender + Marital_status + Age * Gender + Age * Marital_status + Gender * Marital_status + Age * Gender * Marital_status

The significance of Levine's test is under 0.05(Sig =0.006), which suggests that the equal variances assumption is violated. As because the F value is very less(F=2.185), which not harms much. So here it is considering for the 3-way ANOVA. Referred in table -12.

Table -13: Tests of Between-Subjects Effects for H_2 (Sample -2)

Dependent Variable: Happy with the profession							
Source	Type III Sum of Squares	D.f	Mean Square	F	Sig.	Partial Eta Squared	Observed Power[b]
Corrected Model	20.866[a]	18	1.159	1.474	.109	.168	.880
Intercept	379.411	1	379.411	482.518	.000	.786	1.000

Age	8.106	9	.901	1.145	.336	.073	.547
Gender	.101	1	.101	.129	.720	.001	.065
Marital status	5.070	1	5.070	6.448	**.012**	.047	.713
Age * Gender	1.098	3	.366	.466	.707	.011	.142
Age * Marital status	5.913	3	1.971	2.507	.062	.054	.610
Gender * Marital status	2.961	1	2.961	3.765	.054	.028	.487
Age * Gender * Marital status	.000	0	.	.	.	.000	.
Error	103.007	131	.786				
Total	2237.000	150					
Corrected Total	123.873	149					

a. R Squared = .168 (Adjusted R Squared = .054)

b. Computed using alpha = .05

Marital status is only one demographic factor which differentiate over the farmer's happiness in their profession as the F value 6.448 is significant as 0.012 (Table -13).The above said hypothesis is consider not supported. The value is very less so the result is very less applicable for the managerial implication.

4.2.5: Differentiation of happy with the profession over marketing habits

The differentiation of farmers-cum-businessmen's are happy with their profession over marketing habits (How many time come to market sale product, distance of market and amount of sale) is tested by the bellow mentioned hypothesis and the hypothesis is tested by 3-way ANOVA.

H_3: There is significant difference of happiness of farmers-cum-businessmen's over their marketing habits in vegetable market.

Table 14 Levene's Test of Equality of Error Variancesfor H3(Sample -1)

Dependent Variable: Happy with the profession			
F	**df1**	**df2**	**Sig.**
2.028	48	101	.002

Tests the null hypothesis that the error variance of the dependent variable is equal across groups.

a. Design: Intercept + average amount of sales + how many time come to market sale product + distance firm market + average amount of sales * how many time come to market sale product + average amount of sales * distance firm market + how many time come to market sale product * distance firm market + average amount of sales * how many time come to market sale product * distance firm market

The significance of Levine's test is under 0.05(Sig =0.002), which suggests that the equal variances assumption is violated, but the F value is very less(F=2.028) which refers that there is not much harms for the said empirical analysis, referred from table -14.

Table 15 :Tests of Between-Subjects Effects for H3

Dependent Variable: Happy with the profession							
Source	**Type III Sum of Squares**	**df**	**Mean Square**	**F**	**Sig.**	**Partial Eta Squared**	**Observed Power[b]**
Corrected Model	52.067[a]	40	1.101	1.567	.030	.427	.993
Intercept	470.348	1	470.348	669.031	.000	.869	1.000
How many time come to market sale product	4.896	5	.979	1.393	.233	.065	.474
Distance form market	10.463	5	2.093	2.977	**.015**	.128	.841
Average amount of sales	2.004	4	.501	.713	.585	.027	.223
How many time come to market sale product * distance firm market	17.335	13	1.333	1.897	**.039**	.196	.895
How many time come to market sale product * average amount of sales	3.311	7	.473	.673	.695	.045	.277
Distance of firm from market * average amount of sales	1.028	4	.257	.366	.833	.014	.131
How many time come to market for sale product * distance firm of market * average amount of sales	2.245	4	.561	.798	.529	.031	.248
Error	71.006	101	.703				
Total	2237.000	150					
Corrected Total	123.873	149					
a. R Squared = .427 (Adjusted R Squared = .154)							
b. Computed using alpha = .05							

The 'distance to which farmers-cum-businessmen coming to the market' is an individual variable differentiate over the farmers-cum-businessmen's happiness in their profession as the F value 2.977 is significant 0.015. Again, combined effect of 'the number of time coming to the market' and 'the distance of market' is also differentiate over the farmers-cum-businessmen's satisfaction as F value 1.897 is significant 0.039. The above said hypothesis is consider as not supported. The statistical value is less. So, the result can be applicablefor the managerial implication.

4.2.6: Scale reliability for the consumers' data .

To measure the scale reliability of questionnaire, Cronbach's alpha is used. It is the measurement of the internal consistency, which shows the close relativeness of a set of items as a group.

Table 16 Scale Statistics and Reliability statistics (Sample-2)

Mean	Variance	Std. Deviation	N of Items	Cronbach's Alpha	Cronbach's Alpha Based on Standardized Items	N of Items
142.64	325.883	18.052	43	0.846	0.856	43

The scale (questionnaire) is optimally reliable, i,e. 86% as the standardised value of Cronbach's alpha is 0.856 with 43 items excluding 5 demographic variables, referred in table –16.

4.2.7: Exploration of major factors consumer's satisfaction in vegetable market.

To explore the major factors of consumers' satisfaction in the RMC market, the explorative factor analysis is adopted by taking 37 variable.

Table – 17 KMO and Bartlett's Test: Sample adequacy (Sample -2)

Kaiser-Meyer-Olkin Measure of Sampling Adequacy.		**.774**
Bartlett's Test of Sphericity	Approx. Chi-Square	1966.586
	Df	666
	Sig.	.000

KMO measure of sample adequacy is 0.774, which is significant (0.000). So explorative factor analysis is applicable here as referred from the table-17. The significant (0.000) statistics (0.774) of Bartlett's test of Sphericity Justify that the correlation matrix is not an identity matrix. Which shows, that the sample is very appropriate for factor analysis.

Table 18 Total Variance Explained (Sample-2)

Component	Initial Eigenvalues			Extraction Sums of Squared Loadings			Rotation Sums of Squared Loadings		
	Total	% of Variance	Cumulative %	Total	% of Variance	Cumulative %	Total	% of Variance	Cumulative %
1	7.478	20.211	20.211	7.478	20.211	20.211	4.188	11.320	11.320
2	3.739	10.104	30.316	3.739	10.104	30.316	2.556	6.909	18.228
3	2.047	5.533	35.849	2.047	5.533	35.849	2.434	6.579	24.807
4	1.716	4.637	40.486	1.716	4.637	40.486	2.405	6.501	31.308
5	1.461	3.950	44.435	1.461	3.950	44.435	2.047	5.534	36.842
6	1.389	3.754	48.189	1.389	3.754	48.189	2.022	5.466	42.308
7	1.353	3.656	51.845	1.353	3.656	51.845	1.685	4.555	46.862
8	1.240	3.351	55.196	1.240	3.351	55.196	1.665	4.500	51.362
9	1.152	3.113	58.309	1.152	3.113	58.309	1.655	4.472	55.834
10	1.094	2.957	61.266	1.094	2.957	61.266	1.544	4.172	60.007
11	1.046	2.826	64.092	1.046	2.826	64.092	1.512	4.086	**64.092**
12	.999	2.700	66.792						
13	.885	2.391	69.184						
14	.871	2.355	71.538						
15	.850	2.297	73.836						
16	.798	2.156	75.992						
17	.736	1.989	77.980						
13	.683	1.846	79.826						
19	.668	1.805	81.631						

20	.654	1.769	83.400	
21	.609	1.646	85.046	
22	.573	1.548	86.594	
23	.482	1.304	87.898	
24	.479	1.296	89.194	
25	.436	1.178	90.372	
26	.428	1.158	91.530	
27	.412	1.113	92.643	
28	.368	.993	93.636	
29	.354	.957	94.593	
30	.339	.917	95.510	
31	.333	.901	96.411	
32	.267	.722	97.133	
33	.237	.642	97.774	
34	.230	.622	98.396	
35	.220	.595	98.991	
36	.193	.522	99.513	
37	.180	.487	100.000	
Extraction Method: Principal Component Analysis.				

By principal component method and varimax rotation with 15 iteration 11 factors are extracted. The number of variable more than 1 initial eigenvalue are conceder as factors. These 11 factors explain near about 64% variance as cumulative percentage of rotated square loading is 64.092.

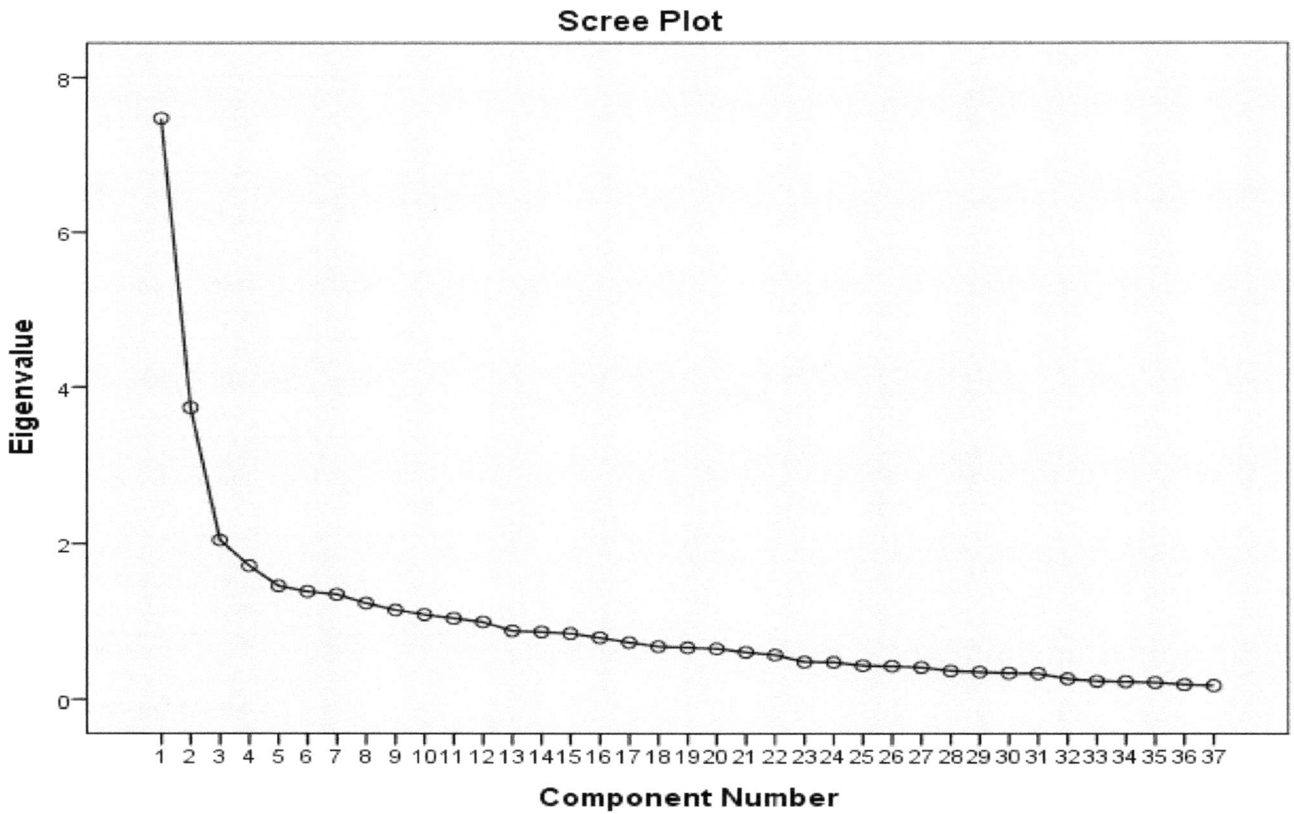

Fig. 4: Scree plot for the consumers' satisfaction

By visualising the scree plot for farmers-cum-businessmen's factors there is sudden elbow mark at the 11 component number in the figure-3 Which, explain the explored factor of the factor analysis

Table 19 Rotated Component Matrix[a](Sample -2)

	Component										
	1	2	3	4	5	6	7	8	9	10	11
Seeing almost the same sellers every day in this market	.820										
Some times the price suddenly increases as farmers-cum sellers sell the vegetables to other whole-sellers and retailers	.723										
Govt. is not taking appropriate steps to regulate properly the RMC markets	.694										
Seasonal products are very cheaper at the pick production season	.672										
Can see the price-increase or decrease of vegetables for the near future	.545	Problems in RMC market									
Like to purchase the vegetable from the farmer directly rather than vegetable merchant	.508										
Having more choices of sellers in this RMC market	.478										
Use to check the quality of vegetable from different seller before purchase the product	.459										
Organised retailer are more systematic than the farmers selling their product at RMC market	.326										

Getting more variety of vegetables from the RMC market		.707			Consumer essentiality
I enjoy purchasing from the RMC market		.681			
RMC market is convenient to purchase the vegetables.		.664			
Now a day vegetables are essential for good health		.586			
Getting fresh and qualitative vegetables at RMC market.		.487			
Getting vegetable at reasonable price from the RMC market		.482			
Getting my product with in my budget prediction			.746		Organised market
Purchasing getting more verity of product from one farmer			.639		
Think the farmers-cum sellers are treating equally to all general customers in this RMC market			.439		
Market place is well structured for the shopping & marketing activities			.386		
Weight measurement in the RMC market seller is quite good				.768	Market structure
Takes little more time to purchase vegetables from this market				.676	
RMC market yard have good parking place				.614	
RMC market yard is well structured for consumer facility				.554	

know about the MSP (minimum supporting price) fixed by the Govt. and accordingly I purchase the products	Customer service	.727			
Getting vegetable at credit from the RMC market		.709			
Use to purchase the product from same seller in most of the time		.480			
Special carry bag for vegetable are provided by the seller of vegetables in RMC market	Product information and quality		.772		
Same quality vegetables are available through-out the day in this RMC market			.504		
Getting appropriate information regarding vegetable from this market.			.478		
I face problem to find out the organic vegetable in RMC market	Market access			.750	
Feel more rush in this market in holidays in comparison to other days.				.537	
This market has more space for shopping activities				.478	
Generally bargain for price with the seller while purchasing the vegetable products in this market	Bargaining power of consumer				.829

Getting directly the firms products from the farmers-cum-sellers which can be store for several days	Freshness of product	.839		
Getting trade discount when I purchase bulk amount of products	Product quality and value		.662	
Getting vegetables in hygienic condition in the RMC market yard			.527	
Farmers-cum business men are behaving properly in RMC market with the general customers	Customer relation			.709
Extraction Method: Principal Component Analysis. Rotation Method: Varimax with Kaiser Normalization.[a]				
a. Rotation converged in 15 iterations.				

By using Principal component analysis and by varimax rotation with Kaiser Normalization, 11 factors are extracted as described in the rotated component matrix table (Table-19) which are named as Problems in RMC market, Consumer essentiality, Organised market, Structure market, Customer service, Product information and quality, Market access, Bargaining power of consumer, Freshness of product, Product quality & value and Customer relation.

4.2.8: Influence of factors of consumers' satisfaction on consumers' positive orientation.

The influence of factors of consumers' satisfaction (11 factors)on consumers' positive orientation is tested by following hypothesis and the hypothesis is tested by the multiple linier regression.

H_4: The customers' satisfaction in vegetable market is significantly influenced by consumers' positive orientation towards the vegetable market.

Originally 'consumer continuity of purchasing vegetable' and 'farmers-cum-businessmen are doing their business happily consumers point of view' are two separate dependent variables. Here the combined effect of above said variable are computed by using transform menu of IBM SPSS software. The new variable is named as 'consumers' positive orientation'.

Table 20 ANOVA[a]: Model fitting for H_4 (sample -2)

Model		Sum of Squares	Df	Mean Square	F	Sig.
1	Regression	208.182	11	18.926	10.295	.000[b]
	Residual	253.691	138	1.838		
	Total	461.873	149			

a. Dependent Variable:

b. Predictors: (Constant), Problems in RMC market, Consumer essentiality, Organised market, Structure market, Customer service, Product information and quality, Market access, Bargaining power of consumer, Freshness of product, Product quality and value, Customer relation

The model, multiple linear regression, is significantly fitting for this analysis as the F statistics (10.295) is significant (p=0.000), referred from the above ANOVA table (table no-20).

Table 21 Model Summary[b] for H_4(sample-2)

Model	R	R Square	Adjusted R Square	Std. Error of the Estimate	Change Statistics					Durbin-Watson
					R Square Change	F Change	df1	df2	Sig. F Change	
1	.671[a]	.451	.407	1.35585	.451	10.295	11	138	.000	1.793

a. Predictors: (Constant), Problems in RMC market, Consumer essentiality, Organised market, Structure market, Customer service, Product information and quality, Market access, Bargaining power of consumer, Freshness of product, Product quality and value, Customer relation

b. Dependent Variable: Consumer's positive orientation

The consumers, satisfying factors influence on consumers' positive orientation by 41% is significant (p=0.000) adjusted R^2 is 0.407. The variables are independent because the variables have no auto co-relation symptoms avelebeble. It is justifying from the Durbin Watson statics (it should very from 1.5 -2.5). In current research it is 1.793 which is justify by the above argument. Referred in table -21.

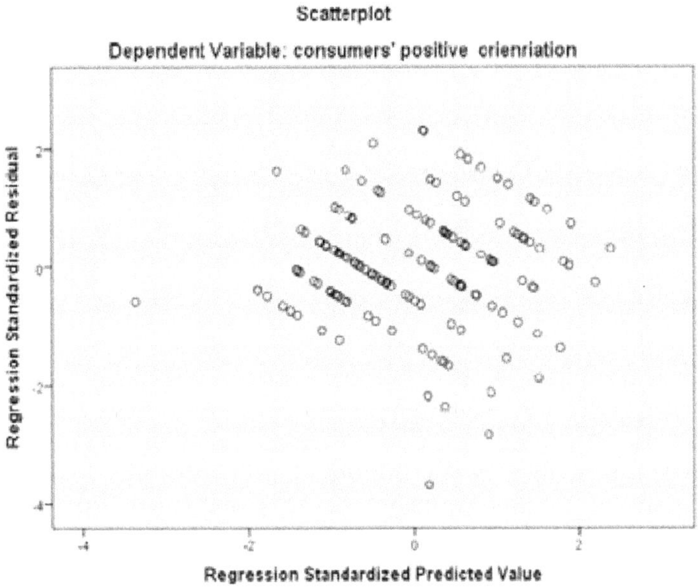

Fig. 5 : Scatterplot, heteroscedacity (Sample-2)

Homoscedasticity assumption is marginally violated as referred in the figure-5 so, the heteroscedasticity symptoms are partially supported here. As the size of the error term not differs large part of values of independent variables hence, the regression model is fitting here.

Table 22 Coefficients[a]:multi-collinearity test for H_4(Sample -2)

		Unstan-dardized Coefficients		Stan-dardized Coeffi-cients			Correlations			Collinearity Statistics	
Model B		**Std. Error**	**Beta**		**T**	**Sig. Zero-or-der**	**Par-tial**	**Part**	**Tol-er-ance**	**VIF**	
1	(Constant)	6.753	.111		61.003	.000					
	Problems in RMC market	.179	.111	.102	1.614	.109	.102	.136	.102	1.000	1.000
	Consumer essentiality	.314	.111	**.178**	2.826	**.005**	.178	.234	.178	1.000	1.000
	Organised market	.366	.111	**.208**	3.294	**.001**	.208	.270	.208	1.000	1.000

	Structure market	.595	.111	**.338**	5.360	**.000**	.338	.415	.338	1.000	1.000
	Customer service	.249	.111	**.142**	2.243	**.026**	.142	.188	.142	1.000	1.000
	Product information and quality	.265	.111	**.151**	2.389	**.018**	.151	.199	.151	1.000	1.000
	Market access	.443	.111	**.252**	3.993	**.000**	.252	.322	.252	1.000	1.000
	Bargaining power of consumer	.073	.111	.042	.661	.510	.042	.056	.042	1.000	1.000
	Freshness of product	.255	.111	**.145**	2.293	**.023**	.145	.192	.145	1.000	1.000
	Product quality and value	-.018	.111	-.010	-.162	.872	-.010	-.014	-.010	1.000	1.000
	Customer relation	.615	.111	**.349**	5.539	**.000**	.349	.426	.349	1.000	1.000

a. Dependent Variable: Consumers positive orientation

Consumer essentiality is an important is contribute 18% (β=0.178 with p=0.005), Organised market contributes21% (β=0.208 with p=0.001), Structure market contributes 34% (β=0.338 with p=0.000), Customer service contribute 14% (β=0.142 with p=0.026), Product information and quality contribute 15 % (β=0.151 with p=0.000), Market access contributes 25% (β=0.252 with p=0.000), Freshness of product contributes 14% (β=0.145 with p=0.000),Customer relation contributes 35% (β=0.349 with p=0.000) to the market efficiency of vegetable marketing. The hypothesis is supported. The value impact is worthy, so it can be taken for the managerial implication.

The resulting regression model can be –

Consumers positive orientation =6.753 + 0.178(Consumer essentiality) + 0.208(Organised market)+ 0..338 (Structure market) + 0.142 (Customer service) + 0.151 (Product information) +0.252(Market access) +0.145(Freshness of product)+0.349(Customer relation)

4.2.9: Interaction of demographic variables with the factors of satisfaction.

To identify the interaction of demographic variables (Age ,Gender , Marital status, Qualification ,Income per month and profession) the factors of satisfaction (11 factors) are tested by following hypothesis and the hypothesis is tested by MANOVA.

H_5: There is significant interaction of demographic factors of consumers with the consumers' positive orientation towards the vegetable shopping in RMC market

Table-23 Box's Test of Equality of Covariance Matrices : for H_5(sample -2)

Box's M	9.780
F	.585
df1	12
df2	775.071
Sig.	.856
Tests the null hypothesis that the observed covariance matrices of the dependent variables are equal across groups.	
a. Design: Intercept + Age ,Gender , Marital status, Qualification ,Income per month and profession and all possible combination of these demographic variables .	

The table-23 shows that the value (9.780) of Box's M is insignificant (p=0.856). So, the results of the box's test shows that the assumption of equal variance relating to dependent variable has not been violated. Hence the result can be produced clearly and can be projected to the population also.

Table -24 Multivariate Test for H_5(Sample -2)

Effect		Value	F	Hypothesis df	Error df	Sig.	Partial Eta Squared	Observed Power[d]
Intercept	Pillai's Trace	.969	651.218[b]	2.000	41.000	.000	.969	1.000
	Wilks' Lambda	.031	651.218[b]	2.000	41.000	.000	.969	1.000
	Hotelling's Trace	31.767	651.218[b]	2.000	41.000	.000	.969	1.000
	Roy's Largest Root	31.767	651.218[b]	2.000	41.000	.000	.969	1.000
Income per month	Pillai's Trace	.452	2.455	10.000	84.000	.013	.226	.924
	Wilks' Lambda	.590	2.474[b]	10.000	82.000	.012	.232	.925
	Hotelling's Trace	.622	2.489	10.000	80.000	.012	.237	.926
	Roy's Largest Root	.469	3.937[c]	5.000	42.000	.005	.319	.915
Profession	Pillai's Trace	.552	3.199	10.000	84.000	.002	.276	.979
	Wilks' Lambda	.509	3.295[b]	10.000	82.000	.001	.287	.982
	Hotelling's Trace	.846	3.384	10.000	80.000	.001	.297	.984
	Roy's Largest Root	.668	5.609[c]	5.000	42.000	.000	.400	.983

Age * Income per month	Pillai's Trace	.339	2.142	8.000	84.000	.040	.169	.819
	Wilks' Lambda	.680	2.177[b]	8.000	82.000	.038	.175	.826
	Hotelling's Trace	.442	2.208	8.000	80.000	.035	.181	.831
	Roy's Largest Root	.364	3.821[c]	4.000	42.000	.010	.267	.857
Gender * profession	Pillai's Trace	.231	2.738	4.000	84.000	.034	.115	.733
	Wilks' Lambda	.780	2.710[b]	4.000	82.000	.036	.117	.727
	Hotelling's Trace	.268	2.680	4.000	80.000	.037	.118	.721
	Roy's Largest Root	.198	4.152[c]	2.000	42.000	.023	.165	.701
Qualification * Income per month	Pillai's Trace	.376	1.943	10.000	84.000	.050	.188	.835
	Wilks' Lambda	.650	1.973[b]	10.000	82.000	.047	.194	.840
	Hotelling's Trace	.500	1.999	10.000	80.000	.044	.200	.845
	Roy's Largest Root	.402	3.379[c]	5.000	42.000	.012	.287	.862
Income per month * profession	Pillai's Trace	.373	2.410	8.000	84.000	.022	.187	.870
	Wilks' Lambda	.659	2.378[b]	8.000	82.000	.023	.188	.864
	Hotelling's Trace	.469	2.344	8.000	80.000	.026	.190	.857
	Roy's Largest Root	.312	3.280[c]	4.000	42.000	.020	.238	.792

a. Design: Intercept and Excluding individual demographic variable, income and profession and combined demographic variable like age & income, gender & profession, qualification & income and income & profession all other variable and the possible combination of variables found to have insignificant interaction with consumer happiness.

b. Exact statistic

c. The statistic is an upper bound on F that yields a lower bound on the significance level.

d. Computed using alpha = .05

The individual effects of income of the consumer is showing significant interaction with 'consumers' positive orientation' as Pillai's Trace value (0.452) is significant (0.013), more is this value between 0 to 1, better is the contribution to the model. Then, by considering value (0.590) of Wilks' Lambda, it is seen that the significance (0.012) is justified. Next, coming to the value (0.622) of Hotelling's Trace, significance (0.012) is proven. Lastly, the value (0.469) of Roy's Largest Root is also seen as significant (0.005). Interpreting in the same manner as above, the individual effect of 'profession' and the combined effects of age*income, gender & profession, qualification & income and income & profession are interacting significantly with the 'consumer's positive orientation in the RMC market'. Thus, the hypothesis is considered to be not supported here. And it cannot be applicable for the managerial implication.

4.2.10: Interaction of purchasing habits with the factors of satisfaction.

To identify the interaction of purchasing habit with the factors of satisfaction of vegetable consumer is tested by the following hypothesis and the hypothesis is tested by MANOVA.

H_6: There is a significant interaction of consumers' purchasing habits with the consumers' positive orientation towards the vegetable shopping in RMC market.

Table 25 Box's Test of Equality of Covariance Matrices: For H_6(sample -2)

Box's M	12.404
F	.509
df1	15
df2	715.995
Sig.	.937
Tests the null hypothesis that the observed covariance matrices of the dependent variables are equal across groups.	
a. Design: Intercept + frequency of purchasing vegetables, distance of firm from market, Products you generally purchasing from RMC market and Average amount purchase as individual variable and all possible combination of these individual variables.	

The table-25 shows that the value (12.404) of Box's M is insignificant (p=0.937). So, the results of the box's test shows that the assumption of equal variance relating to dependent variable has not been violated. Hence the result can be produced clearly and can be projected to the population also.

Table 26 Multivariate Tests: for H_6(Sample-2)

Effect		Value	F	Hypothesis df	Error df	Sig.	Partial Eta Squared	Observed Power[d]
Intercept	Pillai's Trace	.968	671.953[b]	2.000	44.000	.000	.968	1.000
	Wilks' Lambda	.032	671.953[b]	2.000	44.000	.000	.968	1.000
	Hotelling's Trace	30.543	671.953[b]	2.000	44.000	.000	.968	1.000
	Roy's Largest Root	30.543	671.953[b]	2.000	44.000	.000	.968	1.000

frequency of purchasing vegetables* Products generally-purchase from RMC market	Pillai's Trace	.509	2.193	14.000	90.000	.014	.254	.949
	Wilks' Lambda	.549	2.201[b]	14.000	88.000	.014	.259	.949
	Hotelling's Trace	.718	2.206	14.000	86.000	.014	.264	.949
	Roy's Largest Root	.516	3.316[c]	7.000	45.000	.006	.340	.927
products generally-purchase from RMC market * Avrage amount purchase	Pillai's Trace	.487	1.810	16.000	90.000	.042	.243	.915
	Wilks' Lambda	.524	2.096[b]	16.000	88.000	.015	.276	.955
	Hotelling's Trace	.886	2.382	16.000	86.000	.005	.307	.977
	Roy's Largest Root	.862	4.847[c]	8.000	45.000	.000	.463	.994

a. Design: Intercept and frequency of purchasing vegetables, distance of firm from market, products generallypurchase from RMC marketand Average amount purchase as individual variable and all possible combination of these individual variables.

b. Exact statistic

c. The statistic is an upper bound on F that yields a lower bound on the significance level.

d. Computed using alpha = .05

The combined purchase habits 'frequency of purchasing vegetables'* 'products generallypurchase from RMC market' is showing significant interaction with 'consumers' positive orientation' as Pillai's Trace value (0.509) is significant (0.014), more is this value between 0 to 1, better is the contribution to the model. Then, by considering value (0.549) of Wilks' Lambda, it is seen that the significance (0.014) is justified. Next, coming to the value (0.718) of Hotelling's Trace, significance (0.014) is proven. Lastly, the value (0.516) of Roy's Largest Root is also seen as significant (0.006). Interpreting in the same manner as above another combined effects of "products generally purchase from RMC market* 'Average amount purchase' interacting significantly to the consumers' positive orientation in the RMC market. Thus, the hypothesis is considered to be not supported here. And it cannot have a managerial implication.

CHAPTER - 5

Summary Findings

Findings for Objective -1: (To explore the factors of farmer-cum-businessmen satisfaction on vegetable marketing.)

The major factors of farmers-cum-businessmen's satisfaction are contentment of farmers, consumer relationship, market anticipation, market planning in RMC market, farming professionalism, ease of vegetable marketing, marketing cost, proper value of product, convenient selling of vegetable, grading of product, product information & nature and consumer knowledge.

Findings for Objective -2: To extract the factors of customers' satisfaction relating to vegetable market.

The major factors consumer satisfaction by purchasing vegetable in the RMC market are as follows problems in RMC market, consumer essentiality, organised market, structure market, customer service, product information and quality, market access, bargaining power of consumer, freshness of product, product quality & value and customer relation.

Findings for Objective -3: To study the influence of factors farmers-cum-businessmen satisfaction on their happiness towards their current profession.

The farmers-cum-businessmen's satisfying factors are putting impact by 48% on farmers' happiness in their profession relating to near future. All among these satisfying factors, contentment of farmers, farming professionalism and ease of vegetable marketing have significant positive impact on the farmers' happiness. The planners or strategists should emphasize these three factors for the future planning relating to agriculture marketing India. If, it will be issue of maintaining the present level of satisfaction, then the strength above three factors should be maintained.

Findings for Objective -4: To differentiate happiness of farmers-cum-businessmen's over their demographic factor.

Farmers' happiness in their profession is significantly differentiated over marital status of farmers only, not other demographic factors (gender & age). Thus while taking decision or making policy regarding the vegetable marketing, the policy makers should be bothered of making different strategies along the demographic variables.

Findings for Objective -5 To differentiate happiness of farmers-cum-businessmen's over their marketing habits.

The farmers-cum-businessmen happiness differentiated over only one individual factor, 'distance of firm from the resident', not differentiated over other purchasing habits. Hence, the common men's assumption that farmers are reluctant to travel more distance for selling their vegetables, proved to be true by this research result. The Govt. planners or such other strategists should think for more number of systematic RMC market yards to reduce the above difficulty.

Farmers-cum-businessmen happiness also significantly differentiated over combined effect of 'distance of firm' and the 'frequency of coming to market'. It is obvious that distance of firm and frequency of coming to the market have direct relation to each other. As result famers will get more opportunity to sale their vegetable products directly to the consumers, if 'distance' factor will be managed properly by the planners. Also the problem of wastage and undue intervention of middlemen can be solved.

Findings for Objective -6 To study the influence of factors of consumer's satisfaction on consumers' positive orientation towards the vegetable shopping in RMC markets.

Consumers satisfying factors (explored 11 factors) are contributing 41% to the 'consumers' positive orientation' (the new name for mean of two dependent variables). consumer essentiality, organised market, market structure, consumers service product information, market access, freshness of product and consumers' relation are important factors as these factors have significant positive contribution to the model. Rest of the factors are not significant for the above relationship. So, for these factors, the planners or strategists need not be worried.

Further, individual factors like consumer relation (35%), market structure (34%) and market access (25%) are key factors are contributing more to consumers' positive orientation so these factors should take account to make strategy.

Findings for Objective -7 :To judge the interaction of demographic factors with the consumers' positive orientation towards the vegetable shopping in RMC market

The individual demographic factors like income of consumer and the profession interacting with the consumers' positive orientation significantly whereas combined effects of age & income, gender & profession, qualification & income and income & profession are interacting with the "consumer's positive orientation towards the RMC market". It indicates that care should be taken regarding the demographic factors such as income and profession while making any strategic decision to make the vegetable marketing efficient integrating the above said consumers' demographic factors with the farmers' present practice of offering the vegetables in the RMC market.

Findings for Objective -8: To judge the interaction of consumers' purchasing habits with the consumers' positive orientation towards the vegetable shopping in RMC market.

The combined effect of 'frequency of purchasing vegetables' & 'products generally purchase from RMC market' interact significantly with the consumers' positive orientation. Hence the consumers can have a positive thought toward the vegetable market under RMC, which will be reflected through their more frequency of visiting the market and purchase of more verity of product. For this purpose, the planners need to pay attention for availing more products and giving continence to consumers for visiting the market. Another combined effect of factors of purchasing habits, 'products generally purchase from RMC market' &'average amount purchase' interacting significantly with the consumers' positive orientation also must be taken care off in order to make consumers motivated for purchasing vegetables directly from the farmers in the regulated markets.

Overall findings

The farmers-cum-businessmen satisfying factors influence 'farmers-cum-businessmen happiness for the near future with respect to their profession by 48%. On the same time the consumers' satisfying factors influence on consumers' positive orientation by 41% with clear significance of both the values. Thus, any strategic decision regarding the vegetable marketing in the context improved economy, must be taken considering the integral effects of consumers' view point and sellers' view point.

Farmers' happiness in their profession is significantly differentiated only over their marital only, but on the other hand individual demographic factors of consumers like income and the profession interacting with the consumers' positive orientation significantly. Further, combined effects of age & income, gender & profession, qualification & income and income & profession are interacting with the "consumer's positive orientation towards the RMC market". So, there should not to be any bothering about the demographic factor of the farmers-cum-businessmen in the vegetable market because most of the time consumers look towards the product benefits, the demographic factors of sellers. So, marketing strategies of vegetables should be formulated with consideration of consumers' demographic factors irrespective of farmers demographic factors.

The happiness of farmers-cum-businessmen is differentiated with negligible values over the demographic factors and also farmers-cum-businessmen's happiness little differentiated over their purchasing habits. Hence, it is justifying that farmers-cum-businessmen's satisfaction factors (present level of satisfaction) are the key predictors of farmers-cum-businessmen's happiness in their profession (specifically in the future perspective).

CHAPTER - 6

Managerial Implications

The farmer's current knowledge/ information relating to present economy have been effectively related to future growth of the vegetable marketing and hence some strategies are suggested in the study. Some of these important strategies are 'strategic decision regarding the vegetable marketing in the context improved economy, must be taken considering the integral effects of consumers' view point and sellers' view point', 'marketing strategies of vegetables should be formulated with consideration of consumers' demographic factors irrespective of farmers' demographic factors' and 'require more number of systematic RMC market yards to satisfy both the farmers and consumers' .The findings of the study reflect that the Govt. & non-government strategists need to protect the present comfort-ness of farmer-cum-businessmen with respect to the safe and better future of their profession, so that the vegetable production will not become a crisis of the country. Planner or strategist should develop such a strategy in RMC market by which consumers will realize that the RMC market is properly regulated. This can be ensured by the above proposed strategies. Demographic factors of consumers should take into account for policy making in RMC market so that respective consumers will get their desired benefit.

CHAPTER - 7

Conclusion

The vegetable marketing can be effective when the present satisfaction of both the consumers and the farmers-cum-businessmen can be successfully related to the farmers' happiness in the future perspective. Thus, it essential to maintain the present level of farmers-cum-businessmen's satisfaction and consumers positive thinking towards the vegetable marketing. There is a requirement of more and more number of market yards under RMC nearer to the farming area, by which farmers will not reluctant to come market yard. Hence the intervention of middlemen and the wastage of product will be minimised. Communication facility to the RMC market yard need to be develop more market access by the farmers-cum-businessmen. To motivate farmers-cum-businessmen it is a need to make strategy with consideration of market structure, consumer-service, reliable marketing information system, easy market access and good relation among the sealers and consumers. There should not to be bothering about the demographic factors of the farmers-cum-businessmen in the vegetable market because most of the time consumers will look towards the product benefits but not for the demographic factors of sellers as the demographic factors of consumers make variation on their product choice irrespective of farmers' demographic factors in the RMC market. So, the marketing strategies should be formulated by taking care of consumers demographic factors only.

CHAPTER - 8

Limitation and Delimitation

Time limitation made the researchers to go for small sample size 150 from farmers-cum-businessmen's (those farmers are selling their product himself in RMC market yard) and 150 from the consumers those are purchase himself from RMC market but by larger sample size, more approximated results can be obtained.

Only basic software IBM SPSS 23 is used for data analysis, by using more advance technique for data analysis and data collection result of study could be improved.

Respondent for sample-1 of the study are only the farmers, most of them are not literate enough to express their realization fully in the RMC market.

Questionnaire for the sample-1 is administer through the face to face interaction and questionnaire for the sample -2 administrated through the online platform but if, both the sample can be administered in uniform platform better results may have.

Annexure

9.1.1 Questionnaire (Sample -1)

An academic study related to marketing of vegetable products
Responsible Researchers

Tushar Ranjan Sahoo, Research Scholar (MPhil.), Dept. of Business Administration, Sambalpur University

Dr. Saroj Kumar Sahoo. Lecturer, Dept. of Business Administration, Sambalpur University.

Dear Sir/Madam,

It is pleasure for me to select you as a respondent of an intellectual activity, a research work for the partial fulfilment of M.Phil. degree from Sambalpur University. This study is meant for academic purpose only and will not be used for commercial and any such other purposes. I request you to be kind enough to provide your simple responses for the questions and statements given below. It may kindly be noted here that there is no certain answer as completely correct or completely wrong because it is a scientific research work regarding the overall opinion regarding the vegetable product buying, selling, availability, quality and price. So, your free and frank answer will be a great a contribution to this research work, which ultimately serves for the betterment of society. Please see specific instructions that precede every sections of this research instrument, which need to be understood before responding the questions and statements. Research team member(s) will be happy to clarify or assist you, encase you need further understanding about the following questions / statements. If, you need overall results of this study, then please provide your email id at last.

Your cooperation in this regard is highly solicited by me.

Section - A

1. How many time in a month, you come to this market to sale your vegetables. ....
2. What is the distance of your firm/residence from this market...............

5- Completely agree, 4- Agree to a greater extend, 3- moderately agree, 2- little agree, 1- very little agree

Section - B

Sl. no	Statements to be responded on the basis of degree of agree-ness	Degree of Response				
5	Now a day vegetables are essential for good health.	5	4	3	2	1
6	Vegetable are main source of income for me.	5	4	3	2	1
8	Distance of firm location and the market is manageable.	5	4	3	2	1
9	I feel as an entrepreneur in my profession.	5	4	3	2	1
10	Timely I am able to sell my vegetable products.	5	4	3	2	1
12	I chose to sale my product at RMC as first preferences.	5	4	3	2	1
13	I am getting less profit margin due to the more marketing cost.	5	4	3	2	1
15	I am getting easily the necessary information about the market.	5	4	3	2	1
17	I can see the price increase / decrease of vegetables for the near future of this market.	5	4	3	2	1
18	I am visualising the customers' demand for vegetable products in this market.	5	4	3	2	1
19	I think the present production of vegetable products can meet the customers' demand of this market.	5	4	3	2	1
20	I am using proper storing facilities of surplus vegetable at the time of the pick season.	5	4	3	2	1
21	I am getting adequate infrastructure facilities in the RMC market to sell my vegetable products.	5	4	3	2	1
22	I feel customers are satisfied with my behaviour while selling.	5	4	3	2	1
23	I enable to maintain good relation with the consumer.	5	4	3	2	1
24	I am adding some value before selling the vegetables to attract customer like cleaning, putting in their bag properly, & the like.	5	4	3	2	1
25	I am facing problems of loading and unloading of products in this RMC.	5	4	3	2	1
26	I do have proper planning for selling my vegetable products in this RMC market.	5	4	3	2	1

27	I mostly prefer to sell my products to the commission agents as a whole and at a time.	5	4	3	2	1
29	I think little-bit risk is there while selling the vegetable products to the end – users.	5	4	3	2	1
30	I suffer from the marketed surplus of vegetable products in this market.	5	4	3	2	1
31	I am getting proper place in RMC market to sale my products.	5	4	3	2	1
32	Commission agents are affecting negatively the vegetable business in the modern market.	5	4	3	2	1
33	I am getting expected price of vegetables at the high production seasons in this market.	5	4	3	2	1
34	Govt. is providing temperature controlled vehicle for transportation of vegetable.	5	4	3	2	1
36	Unsold product in the market used to sale commission agent.	5	4	3	2	1
40	Wastage of products is reducing my profit in this market.	5	4	3	2	1
41	Consumer has sufficient knowledge about the vegetable products in this market.	5	4	3	2	1
43	I can easily avail the private transportation media for this market.	5	4	3	2	1
44	I do gradating (as high price, medium price & low price) my products before selling in this market.	5	4	3	2	1
45	Perishability nature of vegetable is detracting most of the farmers to produce it for selling purposes.	5	4	3	2	1
47	I got reasonable profit margin from the vegetable product.	5	4	3	2	1
48	I am happy with my profession	5	4	3	2	1
49	I will continue to produce vegetables and sell in RMC market in near future.	5	4	3	2	1

Section – C

Name & Address (optional): ....

Age (in years): Less than 20 ☐ 21-25 ☐ 26-30 ☐ 31-35 ☐ 36-40 ☐
41-45 ☐ 46-50 ☐ 51-55 ☐ 56-60

Above 60 ☐

Gender: Male ☐ Female ☐

Marital Status: Single ☐ Married ☐

Average amount of sales in a day: ....

Email Id: ....

Thanks you.

9.1.2 Questionnaire (Sample -2)

An academic study related to marketing of vegetable products
Responsible Researchers

Tushar Ranjan Sahoo, Research Scholar (MPhil.), Dept. of Business Administration, Sambalpur University

Dr. Saroj Kumar Sahoo. Lecturer, Dept. of Business Administration, Sambalpur University.

Dear Sir/Madam,

It is pleasure for me to select you as a respondent of an intellectual activity, a research work for the partial fulfilment of M.Phil. degree from Sambalpur University. This study is meant for academic purpose only and will not be used for commercial and any such other purposes. I request you to be kind enough to provide your simple responses for the questions and statements given below. It may kindly be noted here that there is no certain answer as completely correct or completely wrong because it is a scientific research work regarding the overall opinion of consumers on the vegetable products' buying, availability, quality and price. So, your free and frank answer will be a great a contribution to this research work, which ultimately serves for the betterment of society. Please see specific instructions that precede every sections of this research instrument, which need to be understood before responding the questions and statements. Research team member(s) will be happy to clarify or assist you, encase you need further understanding about the following questions / statements. If, you need overall results of this study, then please provide your email id at last.

Your cooperation in this regard is highly solicited by me.

Section - A

3. How many time in a month, you come to this market to purchase vegetables. ....
4. What is the distance of your residence from this market....................
5. Which product(s), you are generally purchasing from RMC market : Only vegetable [ ] Vegetable & grocery [ ] All most all products that are available in RMC [ ]

5- Completely agree, 4- Agree to a greater extend, 3- moderately agree, 2- little agree, 1- very little agree

Section-B

Sl. no	**Statements to be responded on the basis of degree of agree-ness**	**Degree of Response**				
4	Now a day vegetables are essential for good health.	5	4	3	2	1
5	I enjoy purchasing from the RMC market.	5	4	3	2	1
6	I am getting more variety of vegetables from the RMC market.	5	4	3	2	1
7	RMC market is convenient to purchase the vegetables.	5	4	3	2	1
8	I am getting vegetable at reasonable price from the RMC market.	5	4	3	2	1
9	I would like to purchase the vegetable from the farmer directly rather than vegetable merchant.	5	4	3	2	1
10	I am getting fresh and qualitative vegetables at RMC market.	5	4	3	2	1
11	I can see the price-increase or decrease of vegetables for the near future.	5	4	3	2	1
12	Farmers-cum business men are behaving properly in RMC market with the general customers.	5	4	3	2	1
13	When purchasing I am getting more verity of product from one farmer.	5	4	3	2	1
14	I am getting appropriate information regarding vegetable from this market.	5	4	3	2	1
15	I think the farmers-cum sellers are treating equally to all general customers in this RMC market.	5	4	3	2	1
16	Seasonal products are very cheaper at the pick production season.	5	4	3	2	1
17	Some times the price suddenly increases as farmers-cum sellers sell the vegetables to other whole-sellers and retailers.	5	4	3	2	1
18	Govt. is not taking appropriate steps to regulate properly the RMC markets.	5	4	3	2	1
19	Organised retailer are more systematic than the farmers selling their product at RMC market.	5	4	3	2	1

20	I know about the MSP (minimum supporting price) fixed by the Govt. and accordingly I purchase the products.	5	4	3	2	1
21	I am getting vegetable at credit from the RMC market.	5	4	3	2	1
22	I use to purchase the product from same seller in most of the time.	5	4	3	2	1
23	I use to check the quality of vegetable from different seller before purchase the product.	5	4	3	2	1
24	I am getting trade discount when I purchase bulk amount of products.	5	4	3	2	1
25	I am getting vegetables in hygienic condition in the RMC market yard.	5	4	3	2	1
26	Same quality vegetables are available through-out the day in this RMC market.	5	4	3	2	1
27	Special carry bag for vegetable are provided by the seller of vegetables in RMC market.	5	4	3	2	1
28	I generally bargain for price with the seller while purchasing the vegetable products in this market.	5	4	3	2	1
29	I have more choices of sellers in this RMC market.	5	4	3	2	1
30	I am seeing almost the same sellers every day in this market.	5	4	3	2	1
31	I am getting directly the firms products from the farmers-cum-sellers which can be store for several days.	5	4	3	2	1
32	RMC market yard have good paring place	5	4	3	2	1
33	RMC market yard is well structured for consumer facility.	5	4	3	2	1
34	Weight measurement in the RMC market seller is quite good.	5	4	3	2	1
35	It takes little more time to purchase vegetables from this market.	5	4	3	2	1
36	I am getting my product with in my budget prediction.	5	4	3	2	1
37	I face problem to find out the organic vegetable in RMC market.	5	4	3	2	1
38	Continue to purchase the vegetable	5	4	3	2	1
39	This market has more space for shopping activities.	5	4	3	2	1
40	I do feel more rush in this market in holidays in comparison to other days.	5	4	3	2	1
41	The market place is well structured for the shopping &marketing activities.	5	4	3	2	1
42	I think farmers-cum-sellers are doing their business happily in this RMC market.	5	4	3	2	1

Section-C

Name & Address (optional):

Age (in years): Less than 20 [ ] 21-25 [ ] 26-30 [ ] 31-35 [ ] 36-40 [ ]
41-45 [ ] 46-50 [ ] 51-55 [ ] 56-60 Above 60 [ ]

Gender: Male [ ] Female [ ]

Marital Status: Single [ ] Married [ ]

Qualification : Below matriculation [ ] +2 Level [ ] +3 Level [ ]
PG. Level [ ]

Income per month (Approximately in Rs) : ....

Profession: ....

Average amount of purchase per month (Approximately in Rs) : ....

Email Id: ....

Thanks you

References

Acharya S.S, & NL, A. (2006). *Agricultural Marketing in India* (4th ed.). Oxford & IBH Publishing Co.Pvt.Ltd.

Agriculture, H. S. D. D. of. (2015). *Horticultural Statistics at a Glance 2015. Oxford University Press YMCA Library Building, 1 Jai Singh Road, New Delhi 110001, India.* Retrieved from http://nhb.gov.in/PDFViwer.aspx?enc=3ZOO8K5CzcdC/Yq6HcdIxC0U1kZZenFuNVXacDLxz28=

Aparna, B., & Hanumanthaiah, C. V. (2013). Organised Retailing of Vegetables : A Farmer , Retailer and Consumer Perspective in India Department of Agricultural Economics , College of Agriculture ,. *American-Eurasian J. Agric. & Environ. Sci, 13*(9), 1258–1265. https://doi.org/10.5829/idosi.aejaes.2013.13.09.11037

B.M. Mehta, & Sonawane, M. (2012). Study of existing vegetable marketing channel with special emphasis on agricultural produce marketing committee (APMC), Dumbhal, Surat. *THE ASIAN JOURNAL OF HORTICULTURE, 7*(1), 78–81.

Chipungahelo, M. S. (2015). Knowledge sharing strategies on traditional vegetables for supporting food security in Kilosa District, Tanzania. *Library Review, 64*(3), 229–247. https://doi.org/10.1108/LR-10-2014-0105

Deliya, M., Thakor, C., & Parmar, B. (2011). A Study on " differentiator in Marketing of fresh fruits and Vegetables from Supply Chain Management Perspective ." *NATIONAL MONTHLY REFEREED JOURNAL OF REASEARCH IN COMMERCE & MANAGEMENT, 1*(1), 40–57.

Dey, S. (2012). Rythu Bazaar: A Study of the Supply Chain of the Farmers' Markets of Andhra Pradesh. *IUP Journal of Operations Management, 11*(3), 43–66. Retrieved from https://search.ebscohost.com/login.aspx?direct=true&db=bth&AN=79827835&site=ehost-live

Geng, X. (2014). Chinese aquatic farmers' participation in modern marketing channels. *British Food Journal, 116*(5), 780–791. https://doi.org/10.1108/BFJ-

01-2013-0010

Global hunger index: India ranks grim 97 of 118 countries on hunger index | India News - Times of India. (2016). Retrieved October 10, 2017, from https://timesofindia.indiatimes.com/india/India-ranks-grim-97-of-118-countries-on-hunger-index/articleshow/54812143.cms?TOI_browsernotification=true

Gunwant, V. K., Hussain, M., R.C, P., Faisal Ali, S. M., & Rana, D. D. (2012). A Comparative Study of Production and Marketing Practices of Vegetables in Nainital and U.S. Nagar Districts of State Uttarakhand, India. *International Journal of Advances in Computing and Information Technology, 1*(6), 569–578. https://doi.org/10.6088/ijacit.12.16006

Gyau, A., & Spiller, A. (2009). Comparison of relationship perception in the international fresh produce business. *Journal of Enterprise Information Management, 22*(4), 441–455. https://doi.org/10.1108/17410390910975040

India ' s Food Sector Cashing in on Forthcoming Opportunities. (2014).

India wastes Rs 44,000 cr of fruits, vegetables and grains annually | india | Hindustan Times. (n.d.). Retrieved June 30, 2017, from http://www.hindustantimes.com/india/india-wastes-rs-44-000-cr-of-fruits-vegetables-and-grains-annually/story-eISAhdFucY9INjVFjkOx7O.html

Joshi, G. (2011). An Analysis of Marketed Surplus and Price Spread of Brinjal in Western Uttar Pradesh. *ASIAN JOURNAL OF MANAGEMENT RESEARCH, 2*(1), 484–490.

Joshi, S. (2012). Role of science and technology for agricultural revival in India. *World Journal of Science, Technology and Sustainable Development, 9*(2), 108–119. https://doi.org/10.1108/20425941211244261

Kahlon, A. . and T. . S. (1983). *Agricultural Price Policy in India*. Allied Publishers Pvt. Ltd., New Delhi.

Karim, R., & Biswas, J. (2016). Value Stream Analysis of Vegetable Supply Chain in Bangladesh: A Case Study. *International Journal of Managing Value and Supply Chains, 7*(2), 41–60. https://doi.org/10.5121/ijmvsc.2016.7205

Kulkarni, K. R. (1964). *Agricultural Marketing in India*. The Co-operators Books Depot, Mumbai.

Kumar, M. P. (2012). a Study on Problems of Marketing Vegetables. *IJRDMS, 6*(1), 241–251.

M. B. Dastagiri, Chand, R., Immanuelraj, T. K., Hanumanthaiah, C. V., Paramsivam, P., Sidhu, R. S., ... Kumar, B. G. (2013). Indian Vegetables: Production Trends, Marketing Efficiency and Export Competitiveness. *American Journal of Agriculture and Forestry, 1*(1), 1. https://doi.org/10.11648/j.ajaf.20130101.11

Man, N., Nawi, N. M., & Ismail, M. M. (2009). an Overview of the Supply Chain

Management of Malaysian. *Journal of Agribusiness Marketing, 2*, 1–18.

Martinez, S. W. (2016). Hedonic analysis of US fresh produce prices at direct-to-consumer sales outlets versus competing retailers. *British Food Journal, 118*(7), 1665–1681. https://doi.org/10.1108/BFJ-05-2015-0183

Mukherjee, S., & Waghmare, A. (2017). India slips 3 notches to 100 on Global Hunger Index; trails N Korea, Iraq | Business Standard News. Retrieved October 26, 2017, from http://www.business-standard.com/article/current-affairs/global-hunger-index-india-slips-3-positions-to-100th-among-119-nations-117101200345_1.html

Munhuweyi, K., Opara, U. L., & Sigge, G. (2016). Postharvest losses of cabbages from retail to consumer and the socio-economic and environmental impacts. *British Food Journal, 118*(2), 286–300. https://doi.org/10.1108/BFJ-08-2014-0280

Nandonde, F. A., & Kuada, J. (2016). Modern food retailing buying behaviour in Africa: the case of Tanzania. *British Food Journal, 118*(5), 1163–1178. https://doi.org/10.1108/BFJ-09-2015-0335

Narasalagi, V. M., & Hegade, G. A. (2013). A Comparative Study on Profitability of Supply Chain Formats in Vegetable Marketing in Karnataka. *Journal of Supply Chain Management Systems, 2*(1), 33–36.

Negi, S., & Anand, N. (2015). Issues and Challenges in the Supply Chain of Fruits & Vegetables Sector in India: A Review. *International Journal of Managing Value and Supply Chains, 6*(2), 47–62. https://doi.org/10.5121/ijmvsc.2015.6205

Niroja, P., Mamoru, I., & Yukio Muto. (2015). Marketing System of Agricultural Cooperatives in Nepal : A Case Studies of Janagarathi Vegetable and Fruit Producer Agriculture , Cooperative. *Review of Integrative Business and Economics Research, 4*(4), 337–348.

Njaya, T. (2014). The Economics of Fruit and Vegetable Marketing by Smallholder Farmers in Murehwa and Mutoko Districts in Zimbabwe. *International Journal of Research in Humanities and Social Sciences, 1*(1), 35–43.

Parwez, S. (2014). Supply Chain Dynamics of Indian Agriculture. *PRODUCTIVITY, 55*(3), 286–295.

Péneau, S., Linke, A., Escher, F., & Nuessli, J. (2009). Freshness of fruits and vegetables: consumer language and perception. *British Food Journal, 111*(3), 243–256. https://doi.org/10.1108/00070700910941453

Pérez Mesa, J. C., & Galdeano-Gómez, E. (2015). Collaborative firms managing perishable products in a complex supply network: an empirical analysis of performance. *Supply Chain Management: An International Journal, 20*(2), 128–138. https://doi.org/10.1108/SCM-06-2014-0185

Pokhrel, D. M. (2010). Comparison of Farm Production and Marketing Cost and Benefit. *Agriculture and Environment, 11*, 10–25.

Pramanik, R., & Prakash, G. (2010). Marketable Surplus and Marketing Efficiency of Vegetables in Indore District : A Micro-Level Study. *The IUP Journal of*

Agricultural Economics, VII(3), 84–94.

Rajagopal, S., & Bansal, I. (2015). Waste disposal of fresh fruits and vegetables. *Management of Environmental Quality: An International Journal, 26*(5), 721–738. https://doi.org/10.1108/MEQ-10-2013-0115

Rajkumar, P., & Jacob, F. (2010). Business Models of Vegetable Retailers in India. *Great Lakes Herald, 4*(1), 31–43. Retrieved from http://www.greatlakes.edu.in/uploads/pdf/Chapter3.pdf

Ruben, R., Boselie, D., & Lu, H. (2007). Vegetables procurement by Asian supermarkets: a transaction cost approach. *Supply Chain Management: An International Journal, 12*(1), 60–68. https://doi.org/10.1108/13598540710724365

Sazzad, P. (2014). Food supply chain management in Indian Agriculture: Issues, opportunities and further research. *African Journal of Business Management, 8*(14), 572–581. https://doi.org/10.5897/AJBM2013.7292

Sheoran A, R. M. (2015). Scope of Supply Chain Management in Fruits and Vegetables in India. *Journal of Food Processing & Technology, 6*(3). https://doi.org/10.4172/2157-7110.1000427

Shukla, M., & Jharkharia, S. (2013). *Agri-fresh produce supply chain management: a state-of-the-art literature review. International Journal of Operations & Production Management* (Vol. 33). https://doi.org/10.1108/01443571311295608

Silva, E., Dong, F., Mitchell, P., & Hendrickson, J. (2015). Impact of marketing channels on perceptions of quality of life and profitability for Wisconsin's organic vegetable farmers. *Renewable Agriculture and Food Systems, 30*(5), 428–438. https://doi.org/10.1017/S1742170514000155

Srimanee, Y., & Routray, J. K. (2012). The fruit and vegetable marketing chains in Thailand: policy impacts and implications. *International Journal of Retail & Distribution Management, 40*(9), 656–675. https://doi.org/10.1108/09590551211255956

Wongprawmas, R., Canavari, M., & Waisarayutt, C. (2015). A multi-stakeholder perspective on the adoption of good agricultural practices in the Thai fresh produce industry. *British Food Journal, 117*(9), 2234–2249. https://doi.org/10.1108/BFJ-08-2014-0300

World Summit on Food Security: The key challenges. (n.d.). Retrieved November 18, 2017, from http://www.fao.org/wsfs/world-summit/wsfs-challenges/en/

Additional charts and graphs

9.3.1 Additional Charts and Graphs (Sample-1)

Table-27.1

In a month coming to market				
		Frequency	Percent	Cumulative Percent
Valid	1-5	14	9.3	9.3
	6-10	36	24.0	33.3
	11-15	44	29.3	62.7
	16-20	39	26.0	88.7
	21-25	13	8.7	97.3
	Above 25	4	2.7	100.0
	Total	150	100.0	

Figure-6.1

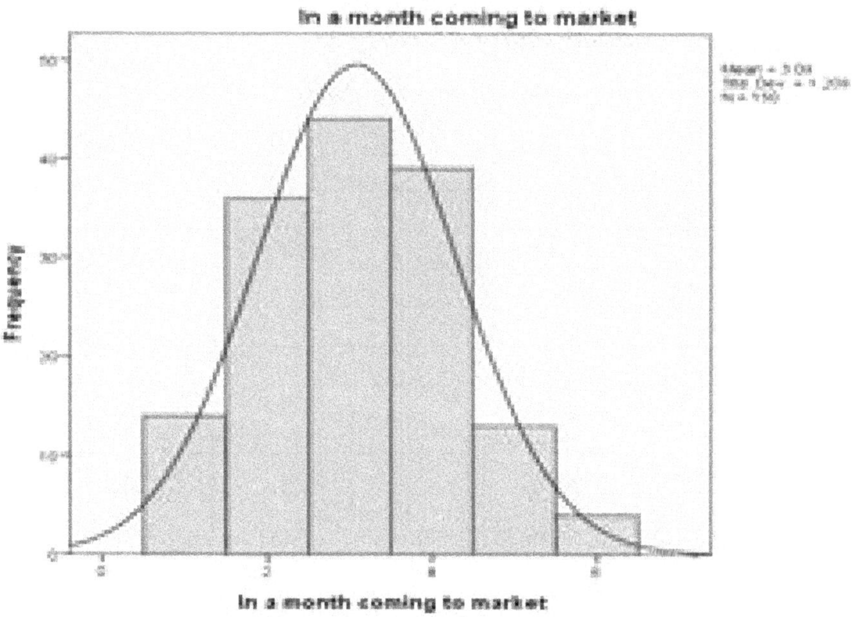

Table-27.2

Distance of the residence from the market				
		Frequency	**Percent**	**Cumulative Percent**
Valid	1-10	54	36.0	36.0
	11-20	40	26.7	62.7
	21-30	30	20.0	82.7
	31-40	9	6.0	88.7
	41-50	14	9.3	98.0
	Above 50	3	2.0	100.0
	Total	150	100.0	

Figure-6.2

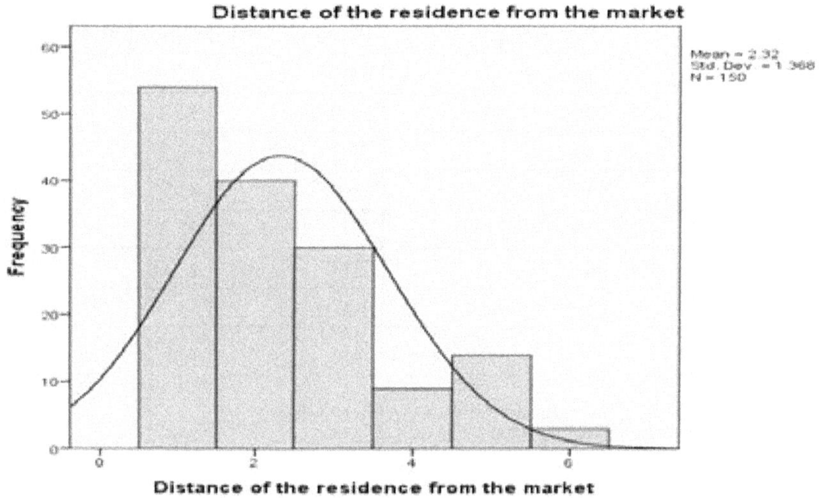

Table-27.3

vegetable is good for health				
		Frequency	Percent	Cumulative Percent
Valid	Very Little Agree	3	2.0	2.0
	Little Agree	14	9.3	11.3
	Moderately Agree	20	13.3	24.7
	Agree to a greater extend	51	34.0	58.7
	Completely Agree	62	41.3	100.0
	Total	150	100.0	

Figure-6.3

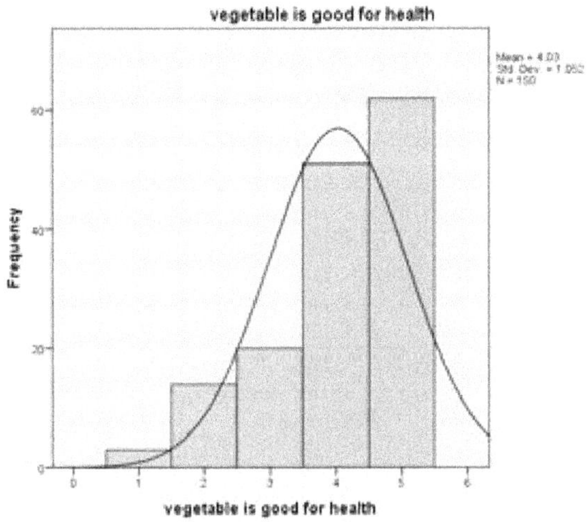

Table-27.4

Vegetable are main source of income				
		Frequency	Percent	Cumulative Percent
Valid	Little Agree	30	20.0	20.0
	Moderately Agree	25	16.7	36.7
	Agree to a greater extend	31	20.7	57.3
	Completely Agree	64	42.7	100.0
	Total	150	100.0	

Figure-6.4

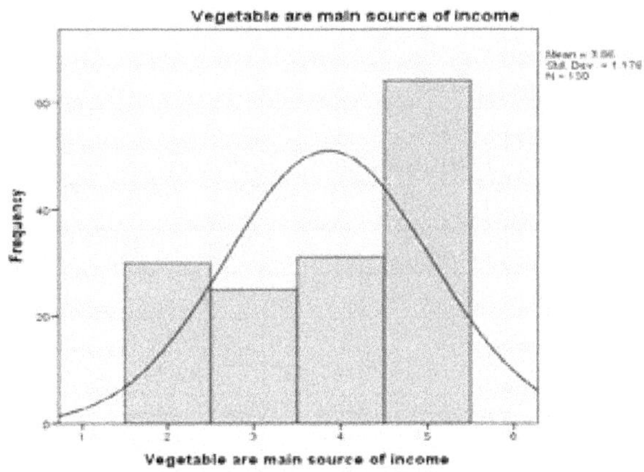

Table-27.5

Distance of firm location and market is manageable				
		Frequency	Percent	Cumulative Percent
Valid	Very Little Agree	4	2.7	2.7
	Little Agree	22	14.7	17.3
	Moderately Agree	22	14.7	32.0
	Agree to a greater extend	56	37.3	69.3
	Completely Agree	46	30.7	100.0
	Total	150	100.0	

Figure-6.5

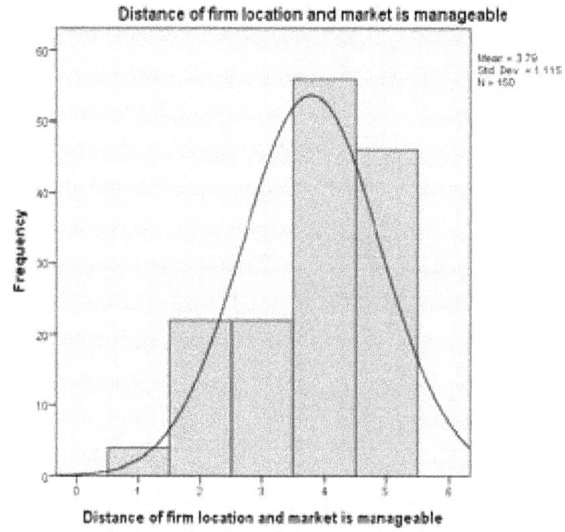

Table-27.6

Fell entrepreneur in profession				
		Frequency	Percent	Cumulative Percent
Valid	Very Little Agree	1	.7	.7
	Little Agree	22	14.7	15.3
	Moderately Agree	32	21.3	36.7
	Agree to a greater extend	33	22.0	58.7
	Completely Agree	62	41.3	100.0
	Total	150	100.0	

Figure-6.6

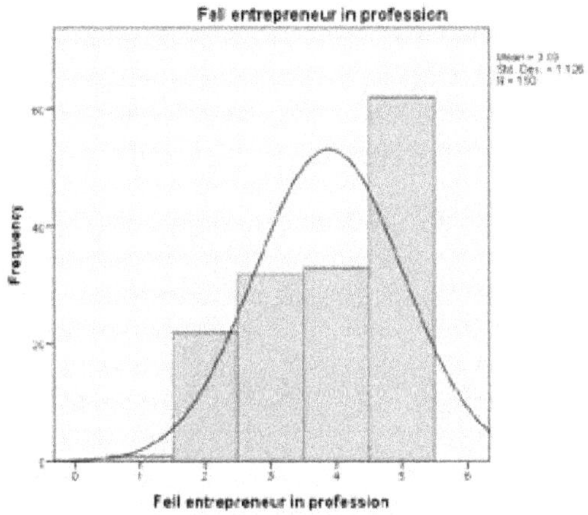

Table-27.7

Timely enable sale vegetable				
		Frequency	Percent	Cumulative Percent
Valid	Very Little Agree	2	1.3	1.3
	Little Agree	12	8.0	9.3
	Moderately Agree	59	39.3	48.7
	Agree to a greater extend	54	36.0	84.7
	Completely Agree	23	15.3	100.0
	Total	150	100.0	

Figure-6.7

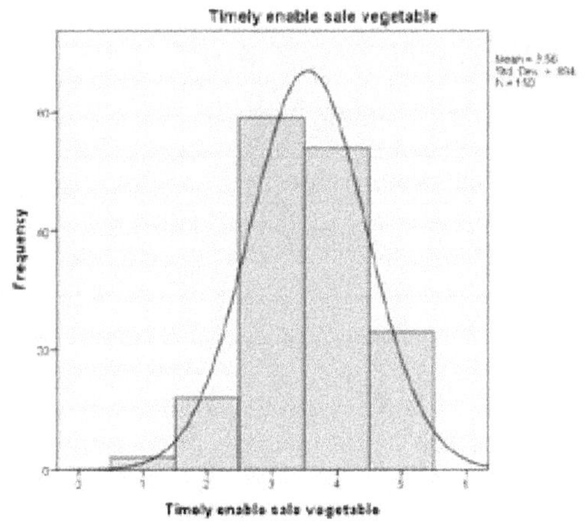

Table-27.8

Chose RMC to sale as first preferences				
		Frequency	Percent	Cumulative Percent
Valid	Very Little Agree	1	.7	.7
	Little Agree	13	8.7	9.3
	Moderately Agree	27	18.0	27.3
	Agree to a greater extend	64	42.7	70.0
	Completely Agree	45	30.0	100.0
	Total	150	100.0	

Figure-6.8

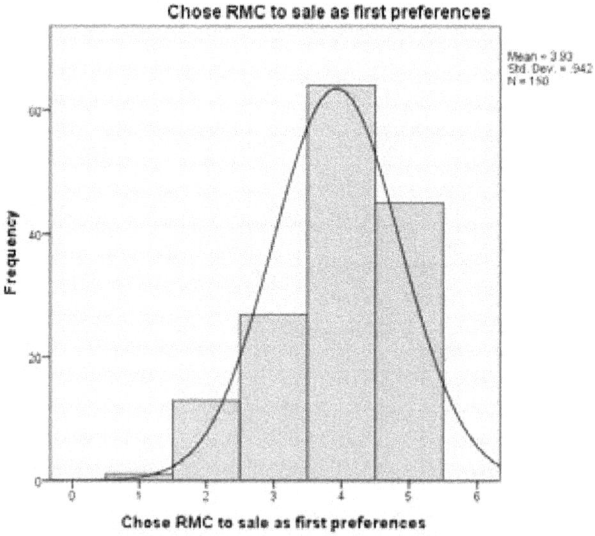

Table-27.9

Less profit margin due to more marketing cost				
		Frequency	Percent	Cumulative Percent
Valid	Very Little Agree	2	1.3	1.3
	Little Agree	13	8.7	10.0
	Moderately Agree	47	31.3	41.3
	Agree to a greater extend	65	43.3	84.7
	Completely Agree	23	15.3	100.0
	Total	150	100.0	

Figure-6.9

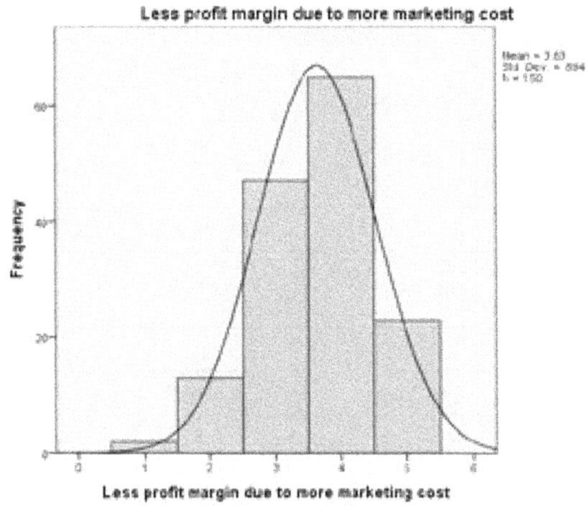

Table-27.10

Getting information about market				
		Frequency	Percent	Cumulative Percent
Valid	Very Little Agree	7	4.7	4.7
	Little Agree	44	29.3	34.0
	Moderately Agree	38	25.3	59.3
	Agree to a greater extend	45	30.0	89.3
	Completely Agree	16	10.7	100.0
	Total	150	100.0	

Figure-6.10

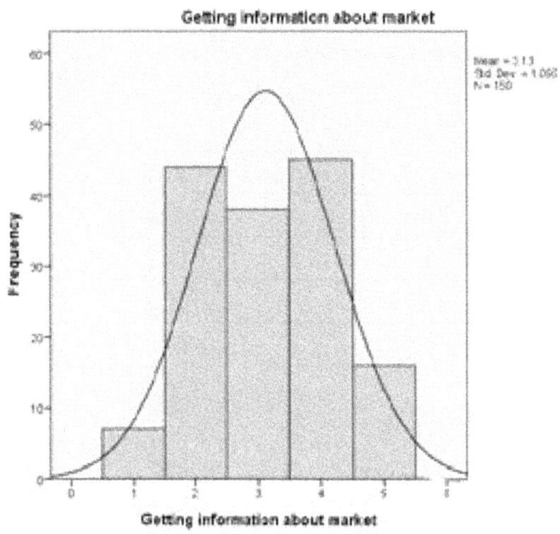

Table-27.11

can see price increase or decrease in near future				
		Frequency	Percent	Cumulative Percent
Valid	Very Little Agree	3	2.0	2.0
	Little Agree	25	16.7	18.7
	Moderately Agree	42	28.0	46.7
	Agree to a greater extend	63	42.0	88.7
	Completely Agree	17	11.3	100.0
	Total	150	100.0	

Figure-6.11

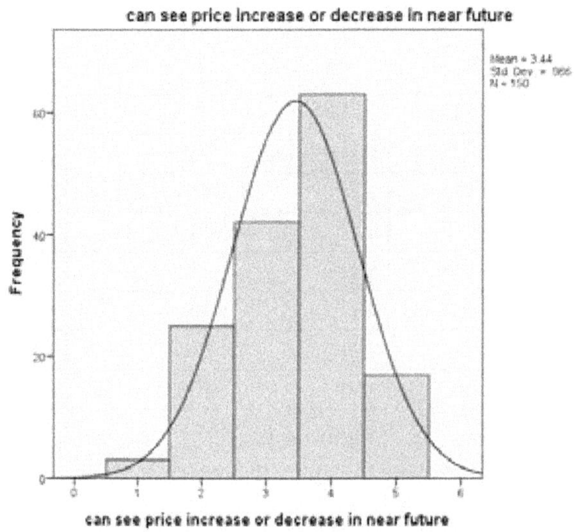

Table-27.12

visualizing the customer demand for product in this market				
		Frequency	Percent	Cumulative Percent
Valid	Very Little Agree	1	.7	.7
	Little Agree	21	14.0	14.7
	Moderately Agree	43	28.7	43.3
	Agree to a greater extend	64	42.7	86.0
	Completely Agree	21	14.0	100.0
	Total	150	100.0	

Figure-6.12

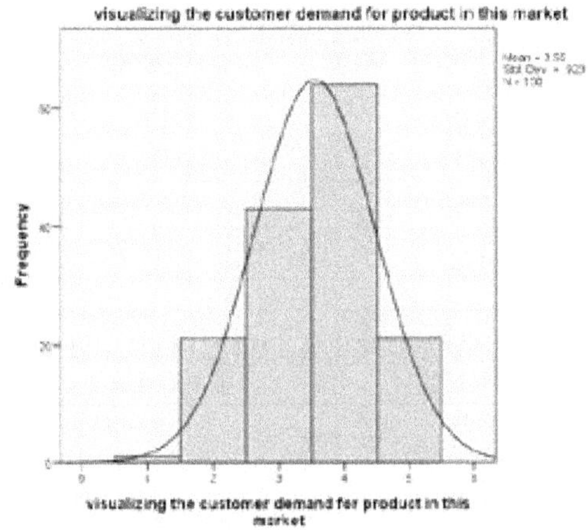

Table-27.13

present production meets the customer demand				
		Frequency	Percent	Cumulative Percent
Valid	Very Little Agree	1	.7	.7
	Little Agree	17	11.3	12.0
	Moderately Agree	29	19.3	31.3
	Agree to a greater extend	65	43.3	74.7
	Completely Agree	38	25.3	100.0
	Total	150	100.0	

Figure-6.13

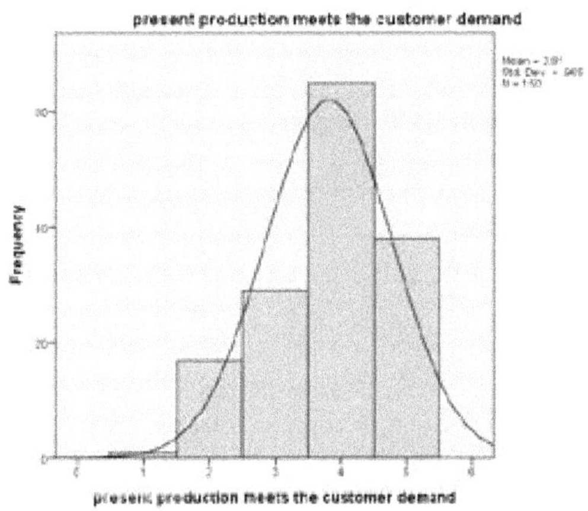

Table-27.14

Proper storing facility of pick season of the vegetable production				
		Frequency	Percent	Cumulative Percent
Valid	Very Little Agree	54	36.0	36.0
	Little Agree	39	26.0	62.0
	Moderately Agree	16	10.7	72.7
	Agree to a greater extend	28	18.7	91.3
	Completely Agree	13	8.7	100.0
	Total	150	100.0	

Figure-6.14

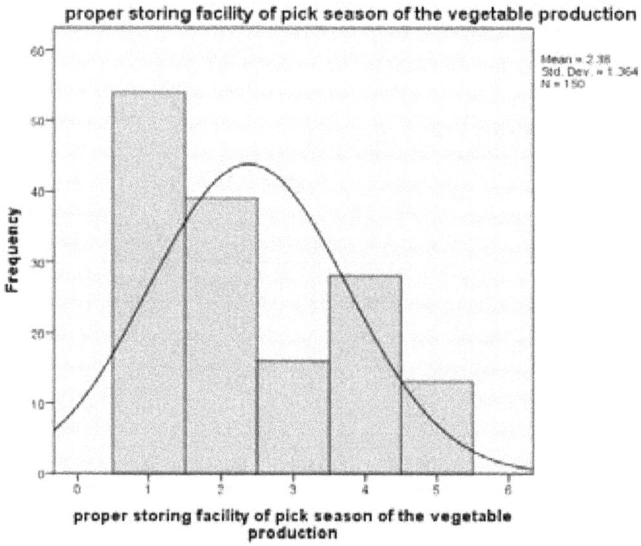

Table-27.15

Getting adequate infrastructure in RMC market				
		Frequency	Percent	Cumulative Percent
Valid	Very Little Agree	6	4.0	4.0
	Little Agree	38	25.3	29.3
	Moderately Agree	54	36.0	65.3
	Agree to a greater extend	24	16.0	81.3
	Completely Agree	28	18.7	100.0
	Total	150	100.0	

Figure-6.15

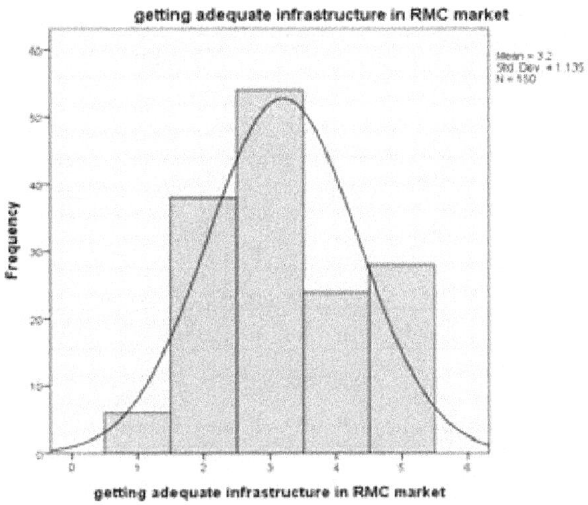

Table-27.16

Fell customer are satisfied with the behaviour				
		Frequency	Percent	Cumulative Percent
Valid	Little Agree	5	3.3	3.3
	Moderately Agree	17	11.3	14.7
	Agree to a greater extend	63	42.0	56.7
	Completely Agree	65	43.3	100.0
	Total	150	100.0	

Figure-6.16

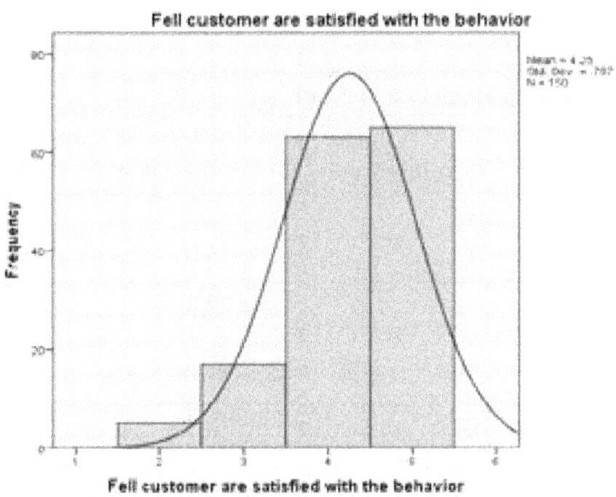

Table-27.17

Enable maintain good relationship with consumer				
		Frequency	Percent	Cumulative Percent
Valid	Very Little Agree	3	2.0	2.0
	Little Agree	4	2.7	4.7
	Moderately Agree	18	12.0	16.7
	Agree to a greater extend	60	40.0	56.7
	Completely Agree	65	43.3	100.0
	Total	150	100.0	

Figure-6.17

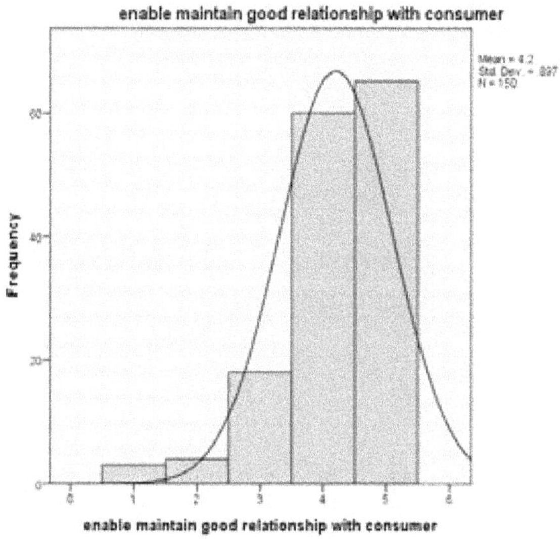

Table-27.18

Adding value before selling to attract consumer				
		Frequency	Percent	Cumulative Percent
Valid	Very Little Agree	11	7.3	7.3
	Little Agree	18	12.0	19.3
	Moderately Agree	33	22.0	41.3
	Agree to a greater extend	65	43.3	84.7
	Completely Agree	23	15.3	100.0
	Total	150	100.0	

Figure-6.18

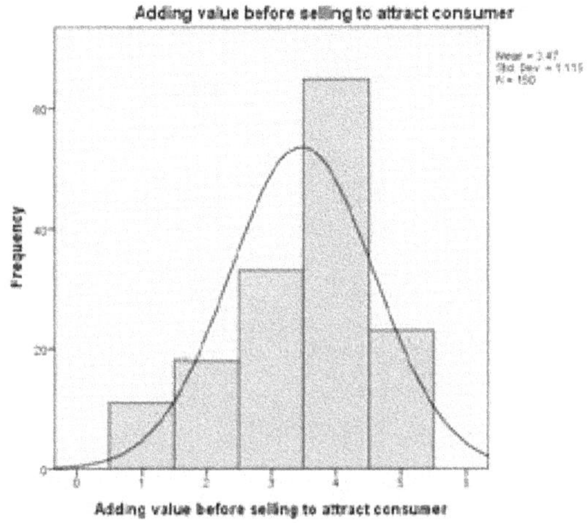

Table-27.19

Facing problem of loading and unloading				
		Frequency	Percent	Cumulative Percent
Valid	Very Little Agree	19	12.7	12.7
	Little Agree	36	24.0	36.7
	Moderately Agree	45	30.0	66.7
	Agree to a greater extend	28	18.7	85.3
	Completely Agree	22	14.7	100.0
	Total	150	100.0	

Figure-6.19

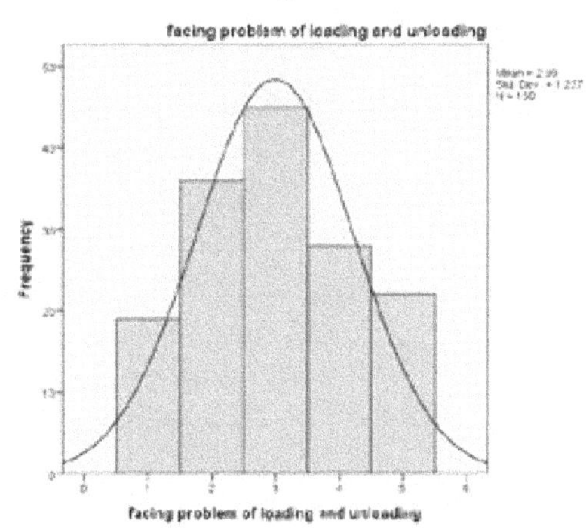

Table-27.20

Having proper planning for selling product in RMC market				
		Frequency	Percent	Cumulative Percent
Valid	Very Little Agree	3	2.0	2.0
	Little Agree	24	16.0	18.0
	Moderately Agree	35	23.3	41.3
	Agree to a greater extend	62	41.3	82.7
	Completely Agree	26	17.3	100.0
	Total	150	100.0	

Figure-6.20

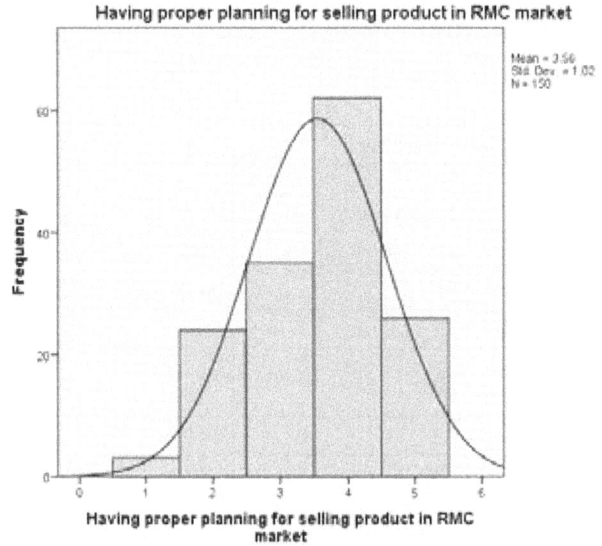

Table-27.21

Prefer sale product as whole at a time				
		Frequency	Percent	Cumulative Percent
Valid	Very Little Agree	11	7.3	7.3
	Little Agree	36	24.0	31.3
	Moderately Agree	43	28.7	60.0
	Agree to a greater extend	29	19.3	79.3
	Completely Agree	31	20.7	100.0
	Total	150	100.0	

Figure-6.21

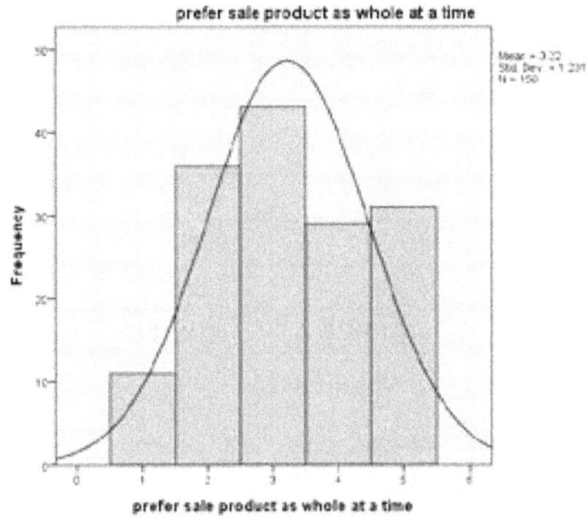

Table-27.22

Think little-bit risk of selling vegetable in the vegetable				
		Frequency	Percent	Cumulative Percent
Valid	Very Little Agree	2	1.3	1.3
	Little Agree	11	7.3	8.7
	Moderately Agree	23	15.3	24.0
	Agree to a greater extend	51	34.0	58.0
	Completely Agree	63	42.0	100.0
	Total	150	100.0	

Figure-6.22

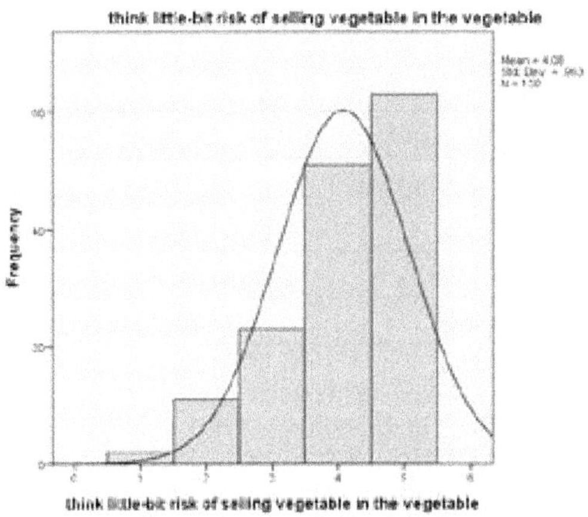

Table-27.23

Suffer marketing surplus in this market				
		Frequency	Percent	Cumulative Percent
Valid	Very Little Agree	13	8.7	8.7
	Little Agree	36	24.0	32.7
	Moderately Agree	50	33.3	66.0
	Agree to a greater extend	38	25.3	91.3
	Completely Agree	13	8.7	100.0
	Total	150	100.0	

Figure-6.23

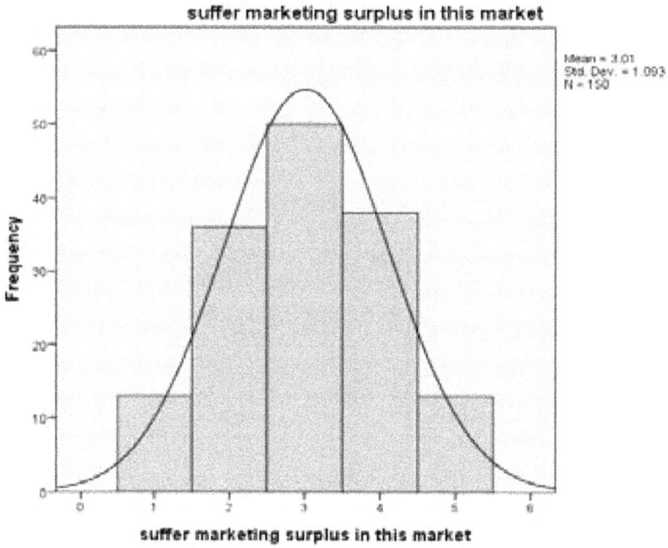

Table-27.24

Getting proper place of selling of vegetable in RMC market				
		Frequency	Percent	Cumulative Percent
Valid	Very Little Agree	20	13.3	13.3
	Little Agree	26	17.3	30.7
	Moderately Agree	33	22.0	52.7
	Agree to a greater extend	35	23.3	76.0
	Completely Agree	36	24.0	100.0
	Total	150	100.0	

Figure-6.24

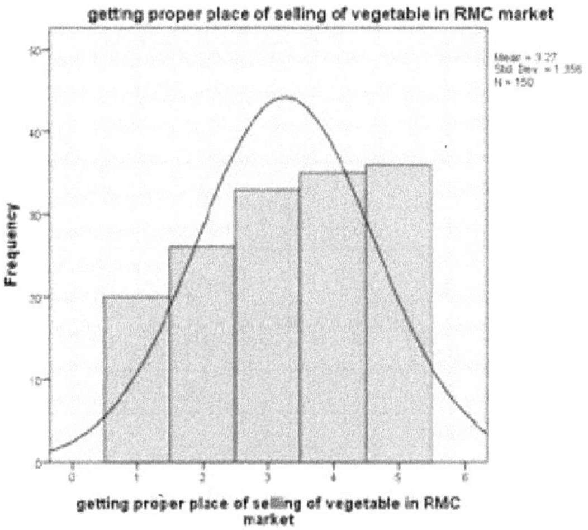

Table-27.25

Commission agents are affect negatively in the business				
		Frequency	Percent	Cumulative Percent
Valid	Very Little Agree	4	2.7	2.7
	Little Agree	36	24.0	26.7
	Moderately Agree	50	33.3	60.0
	Agree to a greater extend	42	28.0	88.0
	Completely Agree	18	12.0	100.0
	Total	150	100.0	

Figure-6.25

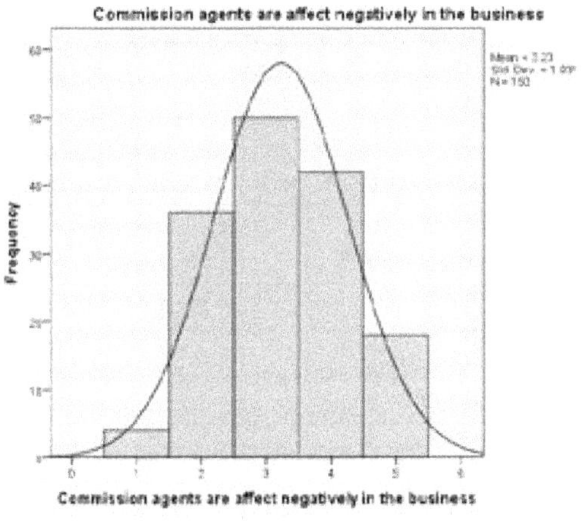

Table-27.26

Getting expected price at the high production seasons				
		Frequency	Percent	Cumulative Percent
Valid	Very Little Agree	23	15.3	15.3
	Little Agree	40	26.7	42.0
	Moderately Agree	49	32.7	74.7
	Agree to a greater extend	27	18.0	92.7
	Completely Agree	11	7.3	100.0
	Total	150	100.0	

Figure-6.26

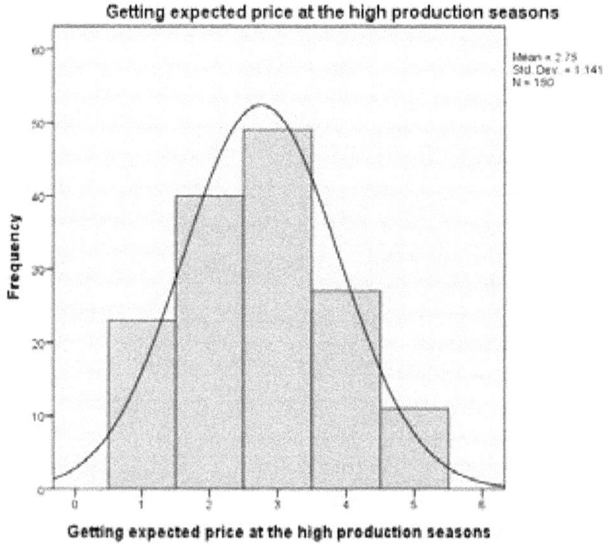

Table-27.27

For transportation of vegetable Government provide temperature control vehicle				
		Frequency	Percent	Cumulative Percent
Valid	Very Little Agree	128	85.3	85.3
	Little Agree	13	8.7	94.0
	Moderately Agree	8	5.3	99.3
	Agree to a greater extend	1	.7	100.0
	Total	150	100.0	

Figure-6.27

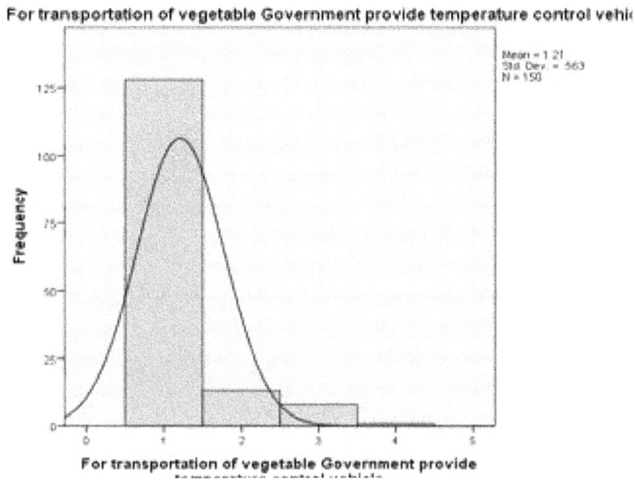

Table-27.28

Prefer to sale unsold product to commission agents				
		Frequency	Percent	Cumulative Percent
Valid	Very Little Agree	20	13.3	13.3
	Little Agree	21	14.0	27.3
	Moderately Agree	32	21.3	48.7
	Agree to a greater extend	50	33.3	82.0
	Completely Agree	27	18.0	100.0
	Total	150	100.0	

Figure-6.28

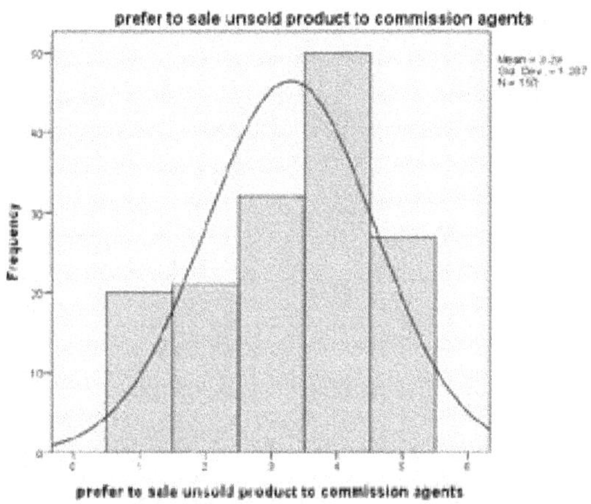

Table-27.29

Wastage of product reduces the profit margin				
		Frequency	Percent	Cumulative Percent
Valid	Very Little Agree	2	1.3	1.3
	Little Agree	17	11.3	12.7
	Moderately Agree	46	30.7	43.3
	Agree to a greater extend	72	48.0	91.3
	Completely Agree	13	8.7	100.0
	Total	150	100.0	

Figure-6.29

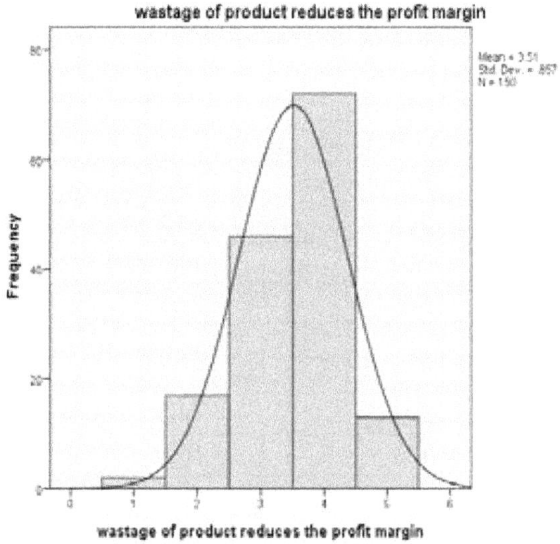

Table-27.30

Consumer have sufficient knowledge about Vegetable				
		Frequency	Percent	Cumulative Percent
Valid	Very Little Agree	2	1.3	1.3
	Little Agree	14	9.3	10.7
	Moderately Agree	29	19.3	30.0
	Agree to a greater extend	78	52.0	82.0
	Completely Agree	27	18.0	100.0
	Total	150	100.0	

Figure-6.30

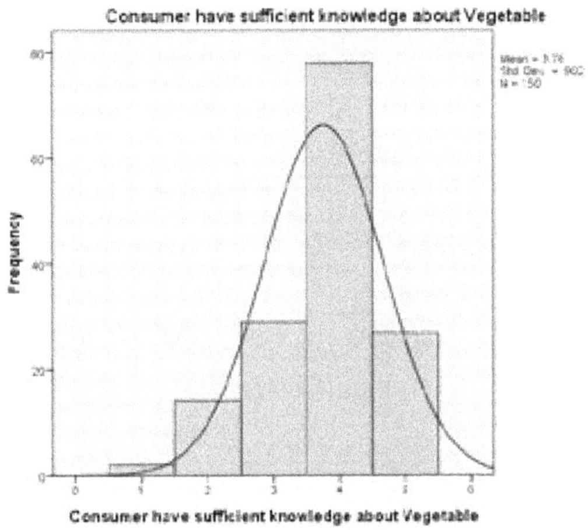

Table-27.31

Easily avail private transportation				
		Frequency	Percent	Cumulative Percent
Valid	Very Little Agree	9	6.0	6.0
	Little Agree	14	9.3	15.3
	Moderately Agree	18	12.0	27.3
	Agree to a greater extend	50	33.3	60.7
	Completely Agree	59	39.3	100.0
	Total	150	100.0	

Figure-6.31

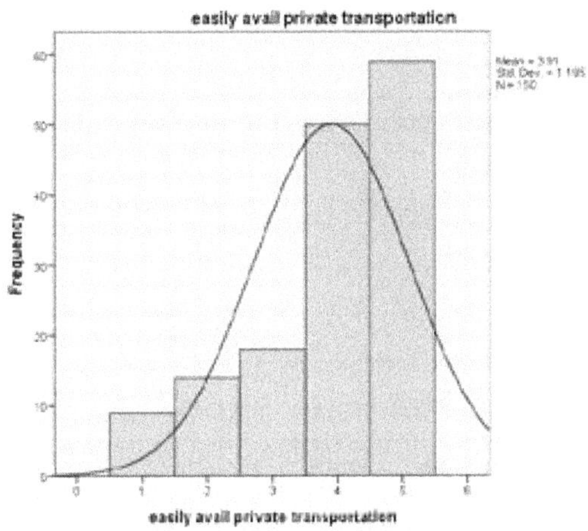

Table-27.32

Grading of product				
		Frequency	Percent	Cumulative Percent
Valid	Very Little Agree	4	2.7	2.7
	Little Agree	10	6.7	9.3
	Moderately Agree	40	26.7	36.0
	Agree to a greater extend	59	39.3	75.3
	Completely Agree	37	24.7	100.0
	Total	150	100.0	

Figure-6.32

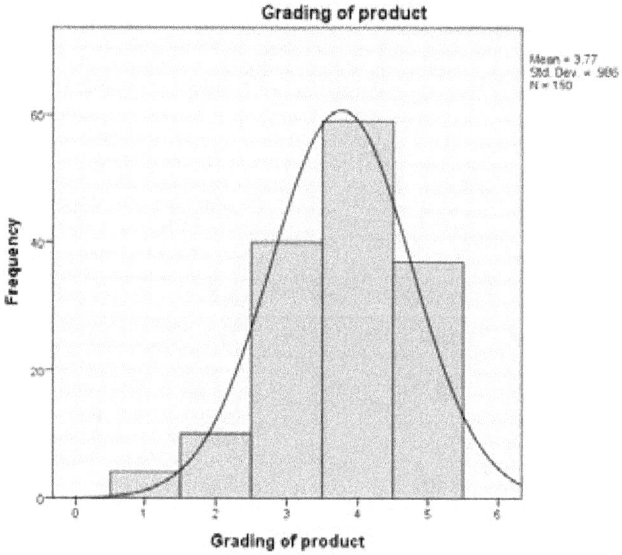

Table-27.32

Perishability of vegetable detracting farmers to produce vegetable for selling propose				
		Frequency	Percent	Cumulative Percent
Valid	Very Little Agree	5	3.3	3.3
	Little Agree	14	9.3	12.7
	Moderately Agree	48	32.0	44.7
	Agree to a greater extend	58	38.7	83.3
	Completely Agree	25	16.7	100.0
	Total	150	100.0	

Figure-6.33

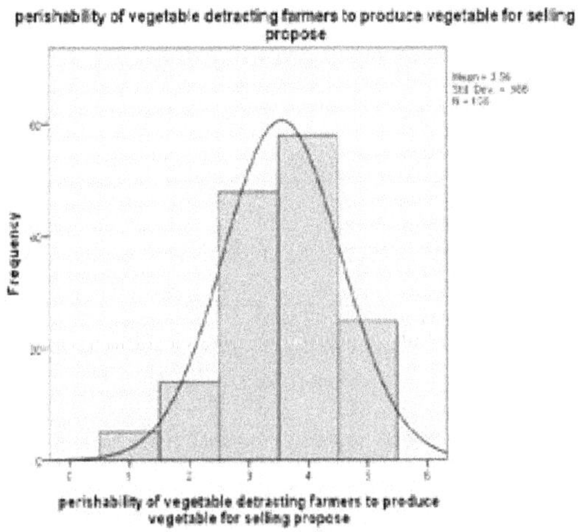

Table-27.34

Getting reasonable profit from vegetable				
		Frequency	Percent	Cumulative Percent
Valid	Very Little Agree	4	2.7	2.7
	Little Agree	22	14.7	17.3
	Moderately Agree	46	30.7	48.0
	Agree to a greater extend	60	40.0	88.0
	Completely Agree	18	12.0	100.0
	Total	150	100.0	

Figure-6.34

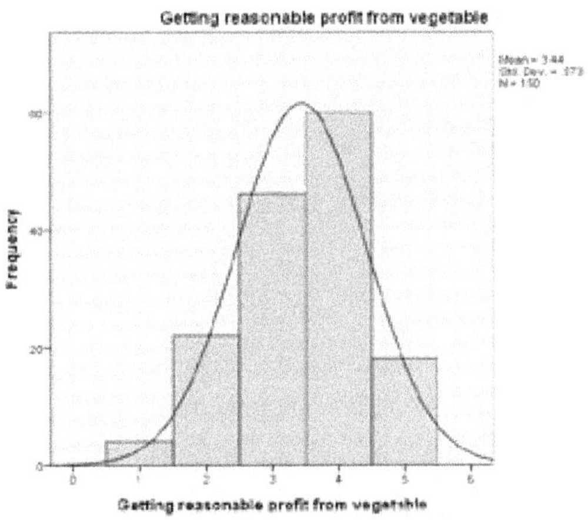

Table-27.35

Happy with the profession				
		Frequency	Percent	Cumulative Percent
Valid	Little Agree	16	10.7	10.7
	Moderately Agree	37	24.7	35.3
	Agree to a greater extend	65	43.3	78.7
	Completely Agree	32	21.3	100.0
	Total	150	100.0	

Figure-6.35

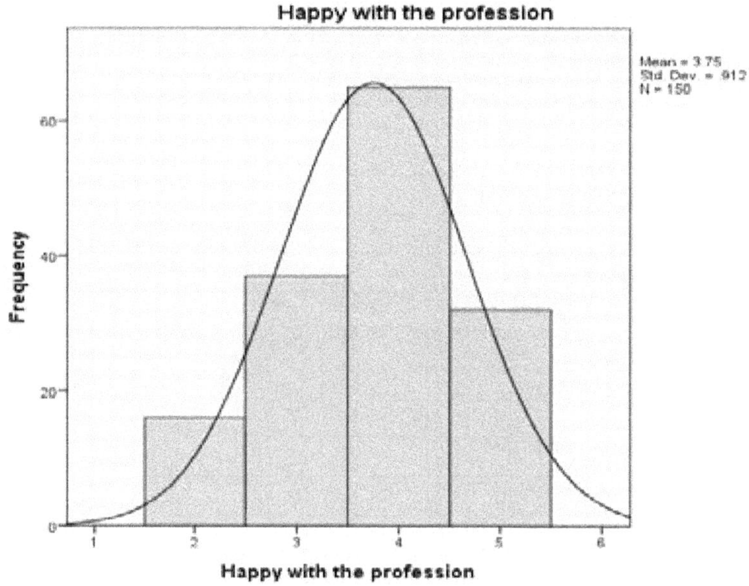

Table-27.36

Continue to sale in RMC market			
	Frequency	Percent	Cumulative Percent

Valid	Very Little Agree	1	.7	.7
	Little Agree	13	8.7	9.3
	Moderately Agree	30	20.0	29.3
	Agree to a greater extend	42	28.0	57.3
	Completely Agree	64	42.7	100.0
	Total	150	100.0	

Figure-6.36

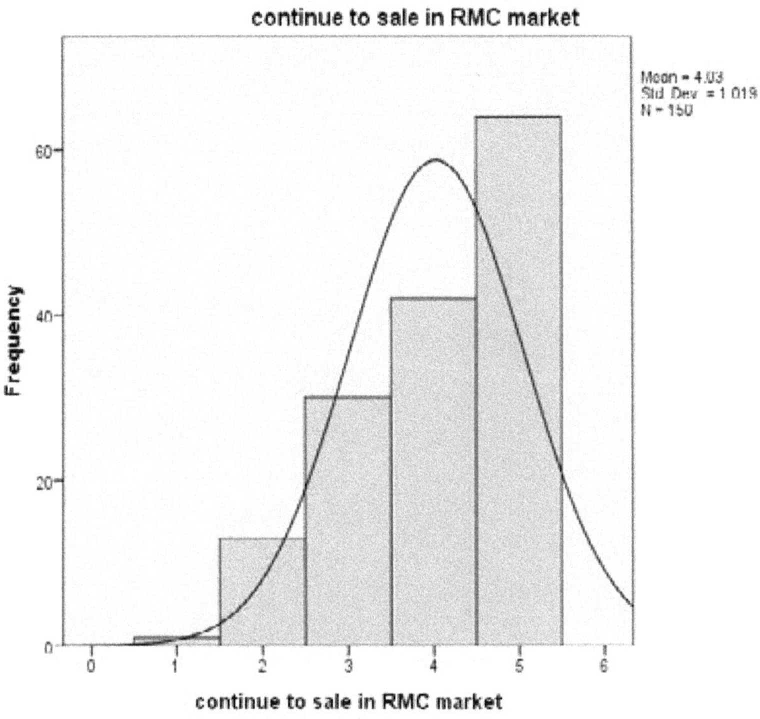

Table-27.37

Age of respondent			
	Frequency	Percent	Cumulative Percent

Valid	less than 20	2	1.3	1.3
	21-25	4	2.7	4.0
	26-30	14	9.3	13.3
	31-35	43	28.7	42.0
	36-40	36	24.0	66.0
	41-45	15	10.0	76.0
	46-50	18	12.0	88.0
	51-55	5	3.3	91.3
	56-60	5	3.3	94.7
	above 60	8	5.3	100.0
	Total	150	100.0	

Figure-6.37

Table-27.38

Gender				
		Frequency	Percent	Cumulative Percent
Valid	male	135	90.0	90.0
	Female	15	10.0	100.0
	Total	150	100.0	

Figure-6.38

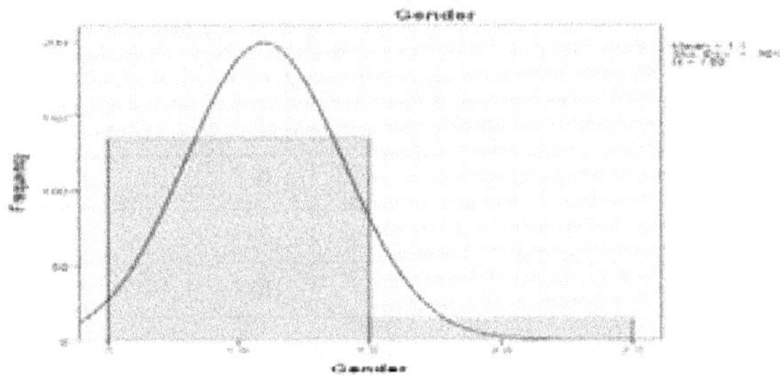

Table-27.39

Marital Status				
		Frequency	Percent	Cumulative Percent
Valid	Single	30	20.0	20.0
	married	120	80.0	100.0
	Total	150	100.0	

Figure-6.39

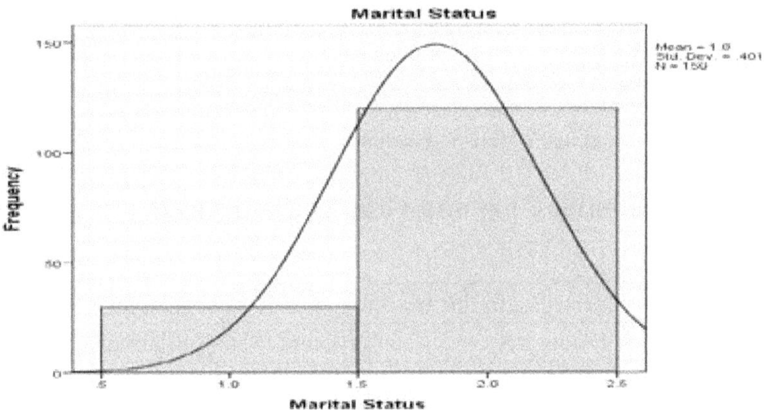

Table-27.40

Average amount of sales			
	Frequency	Percent	Cumulative Percent

Valid	0-2000	101	67.3	67.3
	2001-4000	35	23.3	90.7
	4001-6000	7	4.7	95.3
	6001-8000	3	2.0	97.3
	8001-10000	4	2.7	100.0
	Total	150	100.0	

Figure-6.40

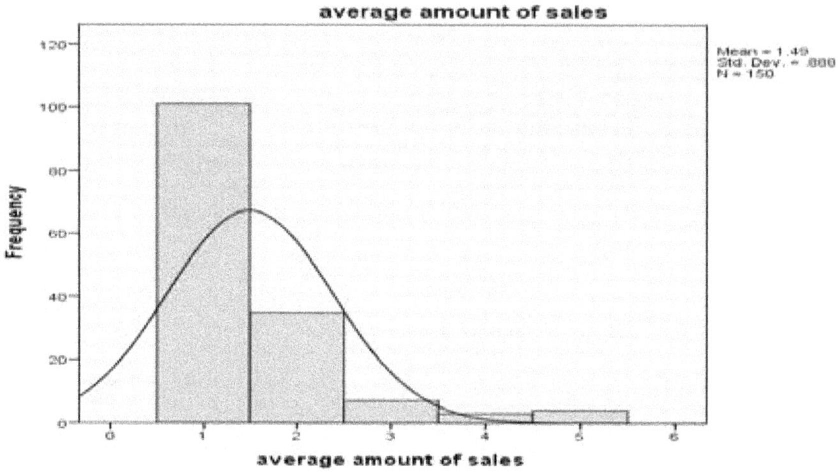

9.3.2 Descriptive Atatistics Chat and Graphs (Sample -2)

Table-28.1

How many time in a month coming to this market to purchase vegetables				
		Frequency	Percent	Cumulative Percent
Valid	1-5	47	31.3	31.3
	6-10	49	32.7	64.0
	11-15	24	16.0	80.0
	16-20	16	10.7	90.7
	21-25	5	3.3	94.0
	Above 25	9	6.0	100.0
	Total	150	100.0	

Figure-7.1

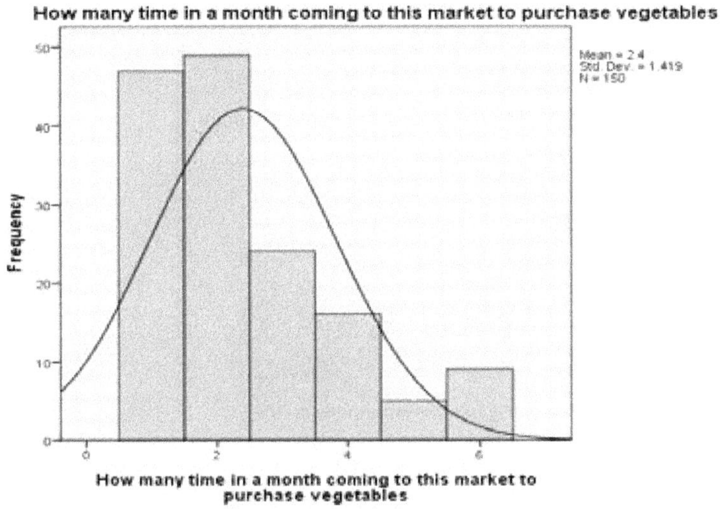

Table-28.2

Distance of the residence from the market				
		Frequency	Percent	Cumulative Percent
Valid	0-2	61	40.7	40.7
	2-4	46	30.7	71.3
	5-6	25	16.7	88.0
	7-8	9	6.0	94.0
	9-10	5	3.3	97.3
	Above 10	4	2.7	100.0
	Total	150	100.0	

Figure-7.2

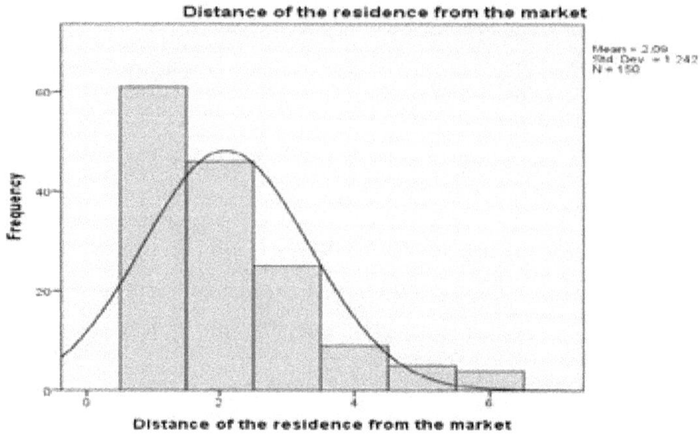

Table-28.3

The product(s), you are generally purchasing from RMC				
		Frequency	Percent	Cumulative Percent
Valid	Only vegetable	42	28.0	28.0
	Vegetable & grocery	60	40.0	68.0
	All most all products that are available in RMC	48	32.0	100.0
	Total	150	100.0	

Figure-7.3

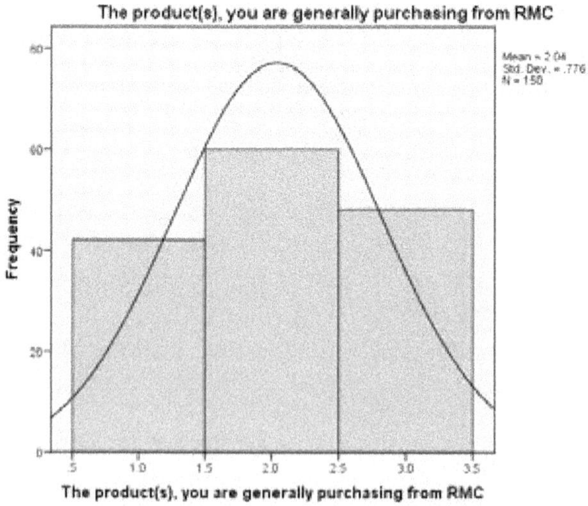

Table-28.4

Now a day vegetables are essential for good health				
		Frequency	Percent	Cumulative Percent
Valid	Little Agree	4	2.7	2.7
	Moderately Agree	33	22.0	24.7
	Agree to a greater extend	31	20.7	45.3
	Completely Agree	82	54.7	100.0
	Total	150	100.0	

Figure-7.4

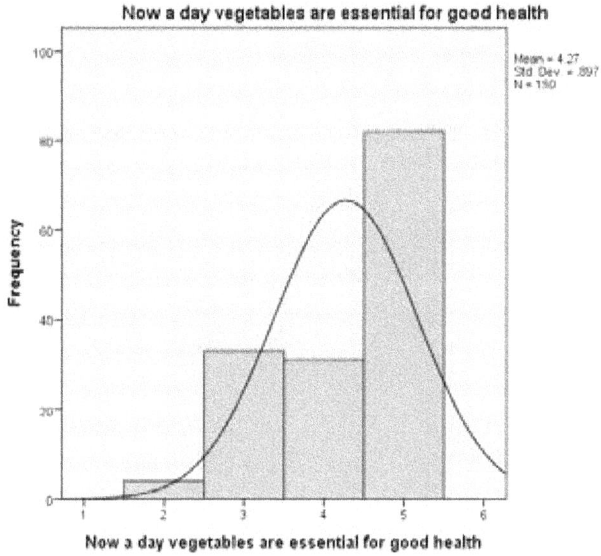

Table-28.5

I enjoy purchasing from the RMC market				
		Frequency	Percent	Cumulative Percent
Valid	Very Little Agree	7	4.7	4.7
	Little Agree	12	8.0	12.7
	Moderately Agree	34	22.7	35.3
	Agree to a greater extend	43	28.7	64.0
	Completely Agree	54	36.0	100.0
	Total	150	100.0	

Figure-7.5

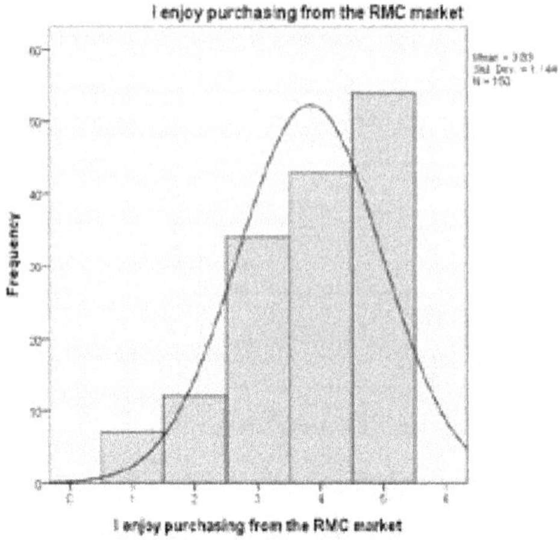

Table-28.6

Getting more variety of vegetables from the RMC market				
		Frequency	Percent	Cumulative Percent
Valid	Very Little Agree	1	.7	.7
	Little Agree	10	6.7	7.3
	Moderately Agree	35	23.3	30.7
	Agree to a greater extend	31	20.7	51.3
	Completely Agree	73	48.7	100.0
	Total	150	100.0	

Figure-7.6

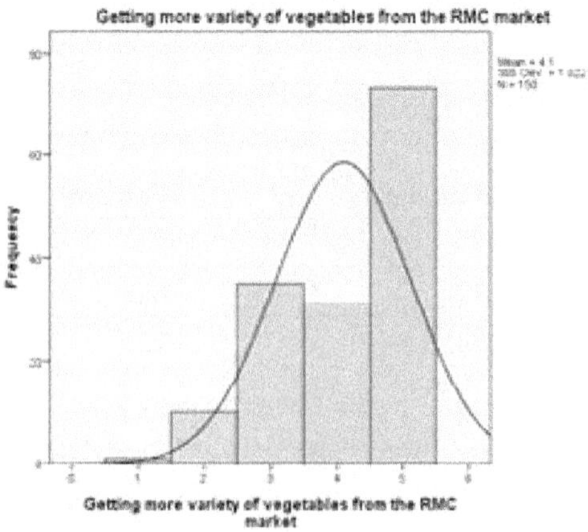

Table-28.7

RMC market is convenient to purchase the vegetables.				
		Frequency	Percent	Cumulative Percent
Valid	Very Little Agree	5	3.3	3.3
	Little Agree	8	5.3	8.7
	Moderately Agree	47	31.3	40.0
	Agree to a greater extend	40	26.7	66.7
	Completely Agree	50	33.3	100.0
	Total	150	100.0	

Figure-7.7

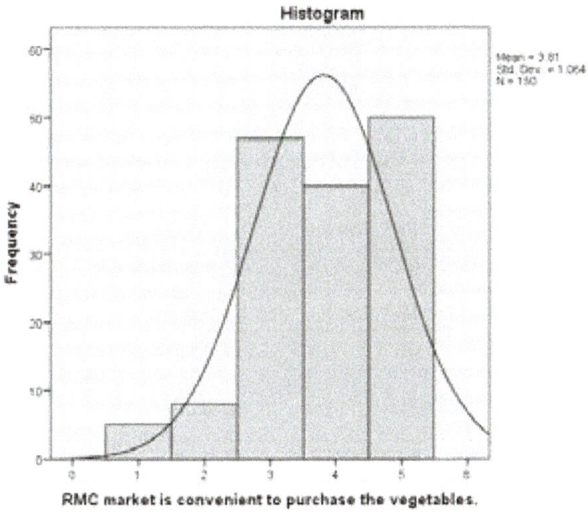

Table-28.8

Getting vegetable at reasonable price from the RMC market				
		Frequency	Percent	Cumulative Percent
Valid	Very Little Agree	6	4.0	4.0
	Little Agree	20	13.3	17.3
	Moderately Agree	52	34.7	52.0
	Agree to a greater extend	33	22.0	74.0
	CCompletely Agree	39	26.0	100.0
	Total	150	100.0	

Figure-7.8

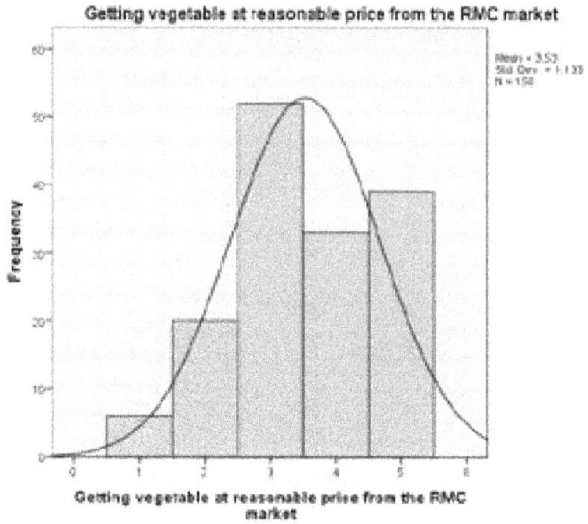

Table-28.9

Like to purchase the vegetable from the farmer directly rather than vegetable merchant				
		Frequency	Percent	Cumulative Percent
Valid	Very Little Agree	3	2.0	2.0
	Little Agree	14	9.3	11.3
	Moderately Agree	46	30.7	42.0
	Agree to a greater extend	31	20.7	62.7
	Completely Agree	56	37.3	100.0
	Total	150	100.0	

Figure-7.9

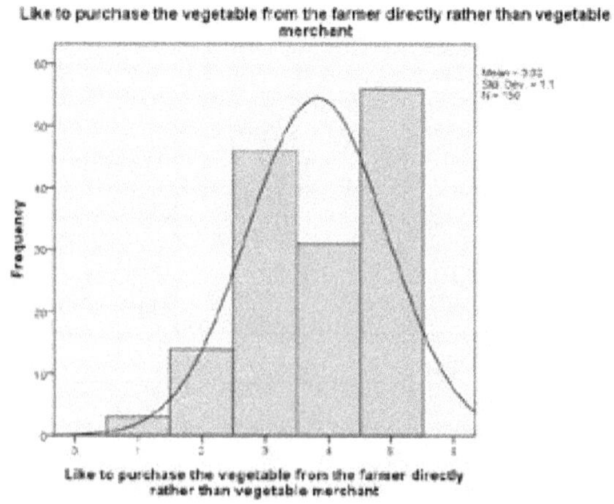

Table-28.10

Getting fresh and qualitative vegetables at RMC market.				
		Frequency	Percent	Cumulative Percent
Valid	Very Little Agree	5	3.3	3.3
	Little Agree	28	18.7	22.0
	Moderately Agree	51	34.0	56.0
	Agree to a greater extend	38	25.3	81.3
	Completely Agree	28	18.7	100.0
	Total	150	100.0	

Figure-7.10

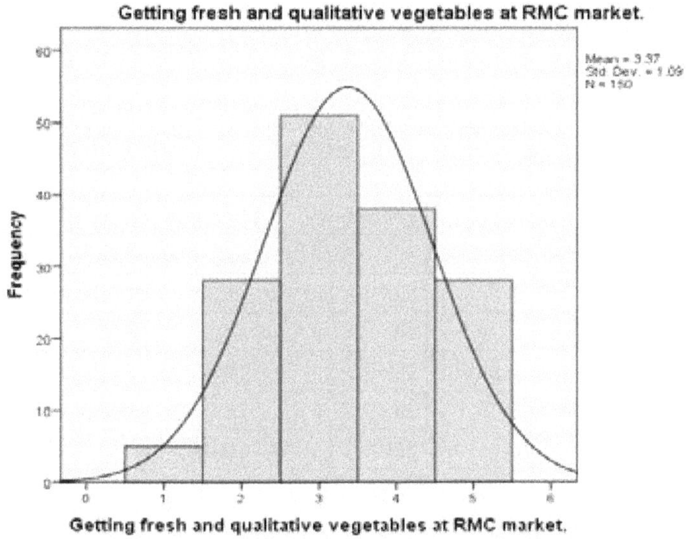

Table-28.11

Can see the price-increase or decrease of vegetables for the near future				
		Frequency	Percent	Cumulative Percent
Valid	Very Little Agree	1	.7	.7
	Little Agree	20	13.3	14.0
	Moderately Agree	52	34.7	48.7
	Agree to a greater extend	34	22.7	71.3
	Completely Agree	43	28.7	100.0
	Total	150	100.0	

Figure-7.11

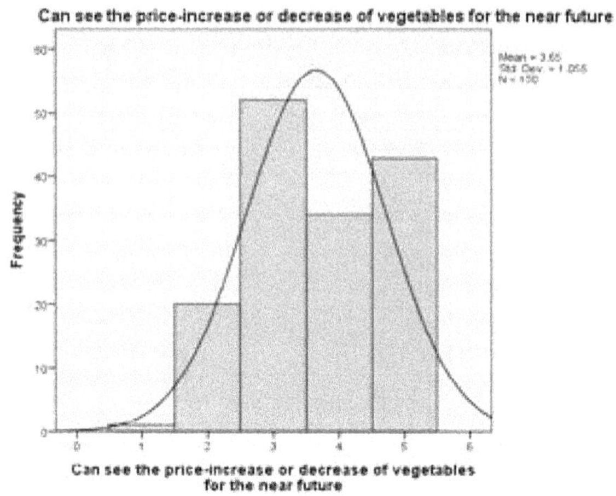

Table-28.12

Farmers-cum business men are behaving properly in RMC market with the general customers				
		Frequency	Percent	Cumulative Percent
Valid	Very Little Agree	9	6.0	6.0
	Little Agree	23	15.3	21.3
	Moderately Agree	65	43.3	64.7
	Agree to a greater extend	29	19.3	84.0
	Completely Agree	24	16.0	100.0
	Total	150	100.0	

Figure-7.12

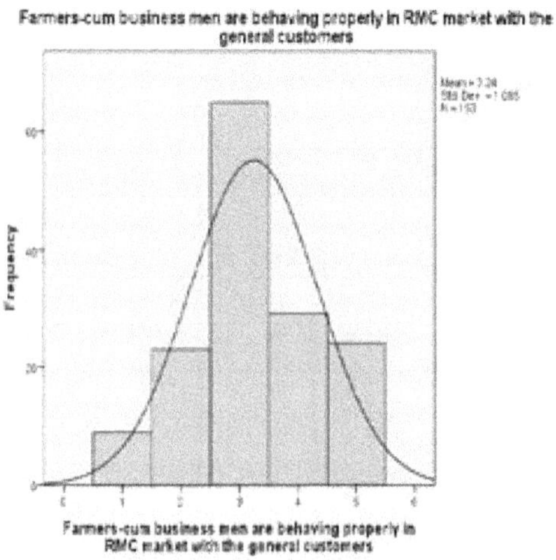

Table-28.13

While Purchasing getting more verity of product from one farmer				
		Frequency	Percent	Cumulative Percent
Valid	Very Little Agree	13	8.7	8.7
	Little Agree	27	18.0	26.7
	Moderately Agree	51	34.0	60.7
	Agree to a greater extend	30	20.0	80.7
	Completely Agree	29	19.3	100.0
	Total	150	100.0	100.0

Figure-7.13

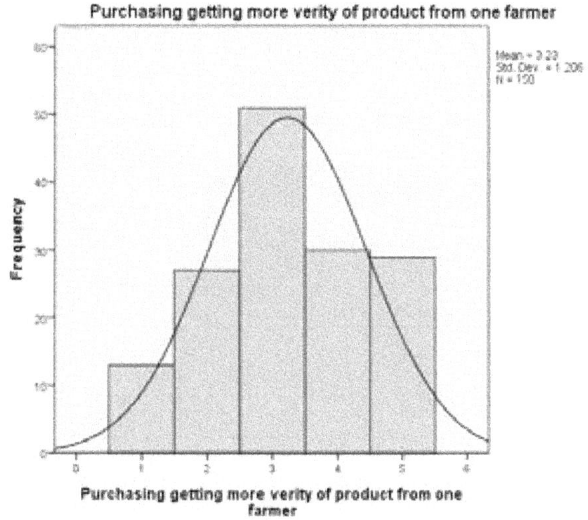

Table-28.14

Getting appropriate information regarding vegetable from this market.				
		Frequency	Percent	Cumulative Percent
Valid	Very Little Agree	9	6.0	6.0
	Little Agree	26	17.3	23.3
	Moderately Agree	60	40.0	63.3
	Agree to a greater extend	30	20.0	83.3
	Completely Agree	25	16.7	100.0
	Total	150	100.0	

Figure-7.14

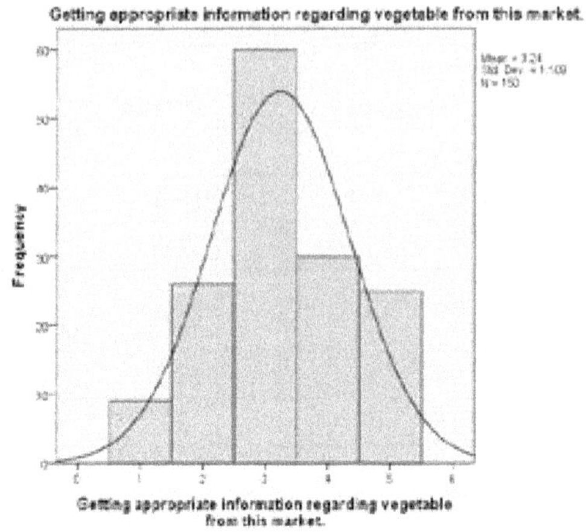

Table-28.15

Think the farmers-cum sellers are treating equally to all general customers in this RMC market				
		Frequency	Percent	Cumulative Percent
Valid	Very Little Agree	9	6.0	6.0
	Little Agree	31	20.7	26.7
	Moderately Agree	53	35.3	62.0
	Agree to a greater extend	28	18.7	80.7
	Completely Agree	29	19.3	100.0
	Total	150	100.0	

Figure-7.15

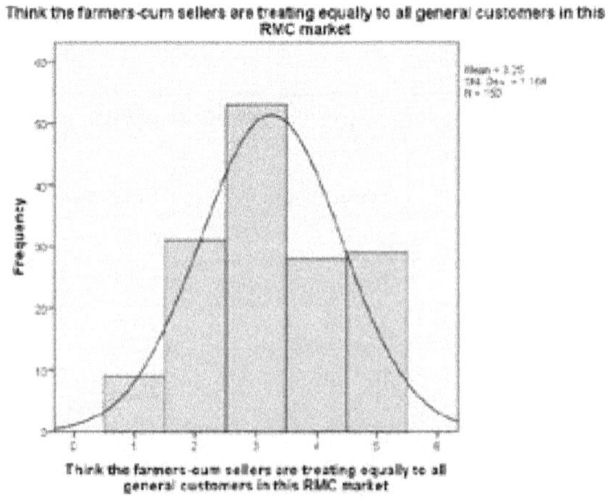

Table-28.16

Seasonal products are very cheaper at the pick production season				
		Frequency	Percent	Cumulative Percent
Valid	Very Little Agree	6	4.0	4.0
	Little Agree	18	12.0	16.0
	Moderately Agree	38	25.3	41.3
	Agree to a greater extend	36	24.0	65.3
	Completely Agree	52	34.7	100.0
	Total	150	100.0	

Figure-7.16

Table-28.17

Some times the price suddenly increases as farmers-cum sellers sell the vegetables to other whole-sellers and retailers				
		Frequency	Percent	Cumulative Percent
Valid	Very Little Agree	1	.7	.7
	Little Agree	19	12.7	13.3
	Moderately Agree	42	28.0	41.3
	Agree to a greater extend	36	24.0	65.3
	Completely Agree	52	34.7	100.0
	Total	150	100.0	

Figure-7.17

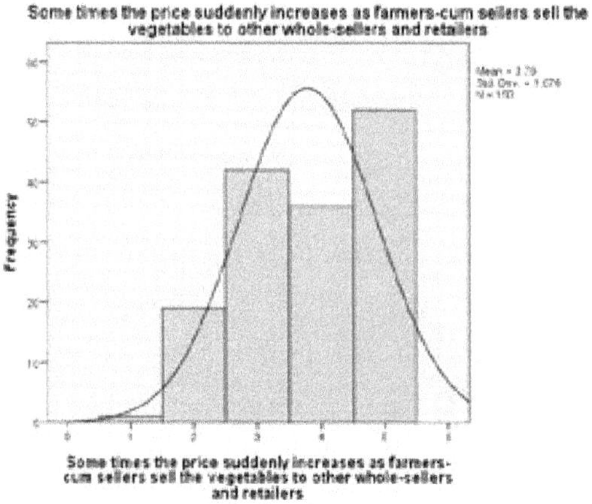

Table-28.18

Govt. is not taking appropriate steps to regulate properly the RMC markets				
		Frequency	Percent	Cumulative Percent
Valid	Little Agree	27	18.0	18.0
	Moderately Agree	41	27.3	45.3
	Agree to a greater extend	30	20.0	65.3
	Completely Agree	52	34.7	100.0
	Total	150	100.0	

Figure-7.18

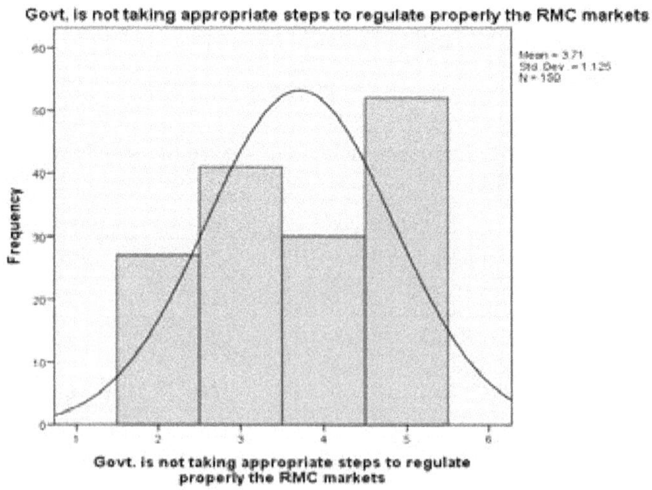

Table-28.19

Organised retailer are more systematic than the farmers selling their product at RMC market				
		Frequency	Percent	Cumulative Percent
Valid	Very Little Agree	6	4.0	4.0
	Little Agree	28	18.7	22.7
	Moderately Agree	47	31.3	54.0
	Agree to a greater extend	39	26.0	80.0
	Completely Agree	30	20.0	100.0
	Total	150	100.0	

Figure-7.19

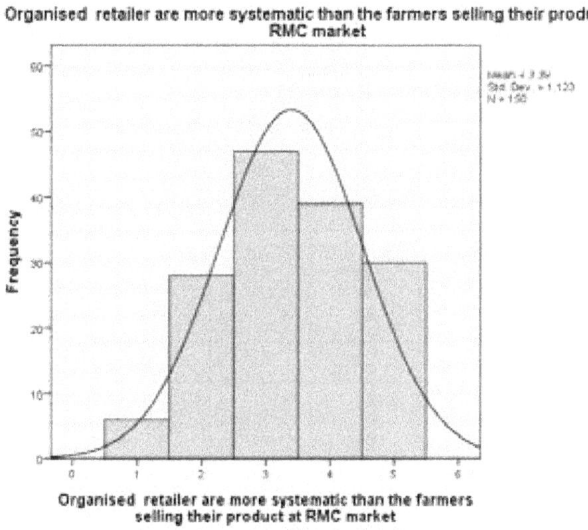

Table-28.20

Know about the MSP (minimum supporting price) fixed by the Govt. and accordingly I purchase the products				
		Frequency	Percent	Cumulative Percent
Valid	Very Little Agree	20	13.3	13.3
	Little Agree	43	28.7	42.0
	Moderately Agree	51	34.0	76.0
	Agree to a greater extend	24	16.0	92.0
	Completely Agree	12	8.0	100.0
	Total	150	100.0	

Figure-7.20

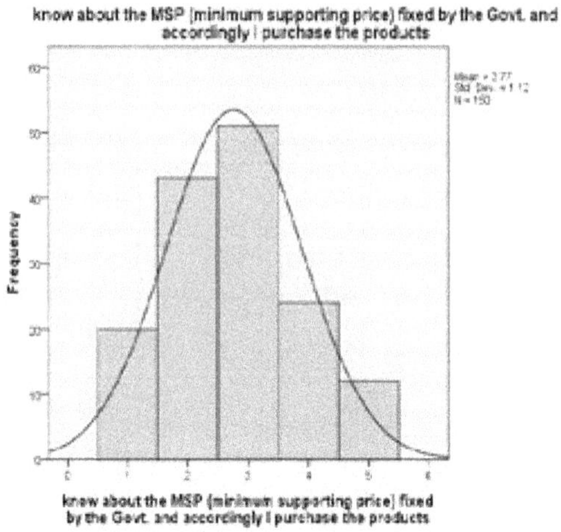

Table-28.21

Getting vegetable at credit from the RMC market				
		Frequency	Percent	Cumulative Percent
Valid	Very Little Agree	46	30.7	30.7
	Little Agree	32	21.3	52.0
	Moderately Agree	38	25.3	77.3
	Agree to a greater extend	20	13.3	90.7
	Completely Agree	14	9.3	100.0
	Total	150	100.0	

Figure-7.21

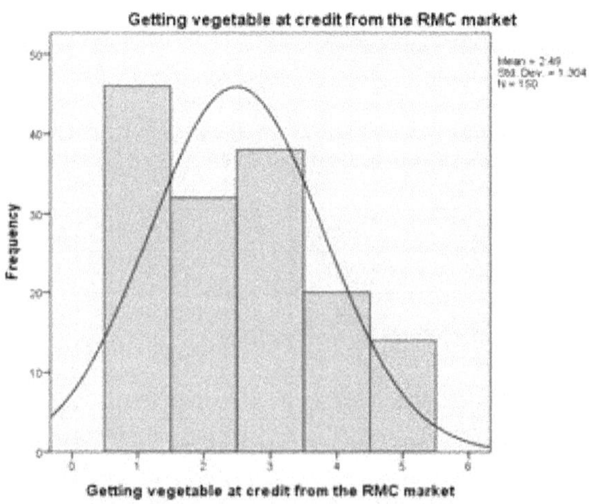

Table-28.22

Use to purchase the product from same seller in most of the time				
		Frequency	Percent	Cumulative Percent
Valid	Very Little Agree	5	3.3	3.3
	Little Agree	32	21.3	24.7
	Moderately Agree	46	30.7	55.3
	Agree to a greater extend	39	26.0	81.3
	Completely Agree	28	18.7	100.0
	Total	150	100.0	

Figure-7.22

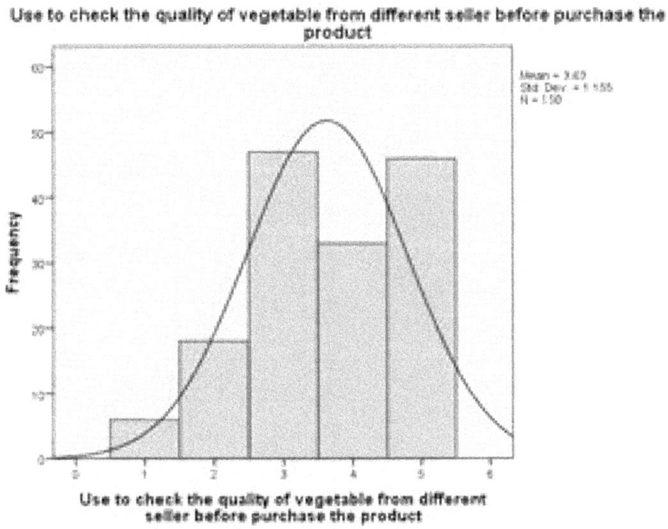

Table-28.23

Use to check the quality of vegetable from different seller before purchase the product				
		Frequency	Percent	Cumulative Percent
Valid	Very Little Agree	6	4.0	4.0
	Little Agree	18	12.0	16.0
	Moderately Agree	47	31.3	47.3
	Agree to a greater extend	33	22.0	69.3
	Completely Agree	46	30.7	100.0
	Total	150	100.0	

Figure-7.23

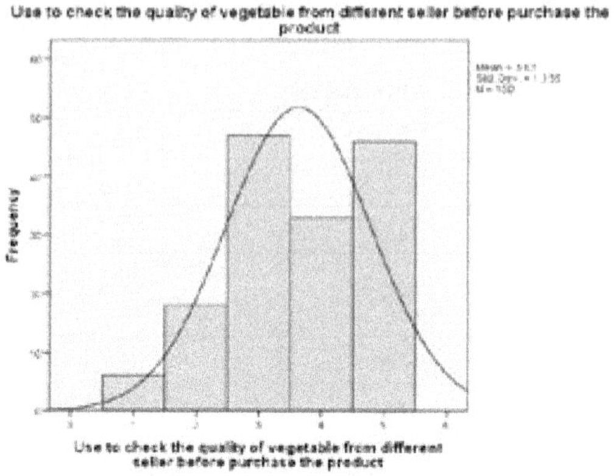

Table-28.24

Getting trade discount when I purchase bulk amount of products				
		Frequency	Percent	Cumulative Percent
Valid	Very Little Agree	5	3.3	3.3
	Little Agree	34	22.7	26.0
	Moderately Agree	46	30.7	56.7
	Agree to a greater extend	30	20.0	76.7
	Completely Agree	35	23.3	100.0
	Total	150	100.0	

Figure-7.24

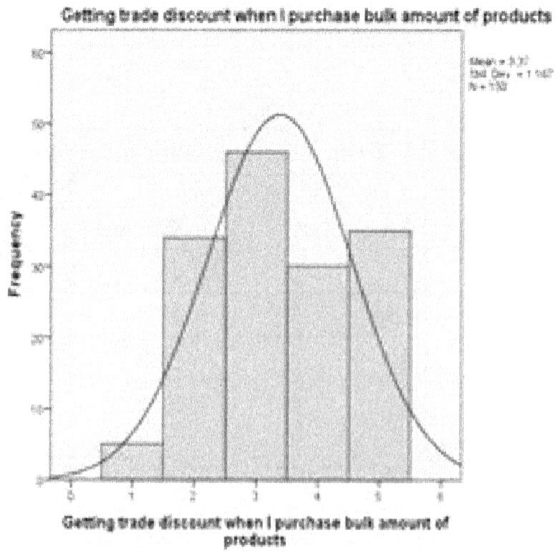

Table-28.25

Getting vegetables in hygienic condition in the RMC market yard				
		Frequency	Percent	Cumulative Percent
Valid	Very Little Agree	11	7.3	7.3
	Little Agree	40	26.7	34.0
	Moderately Agree	46	30.7	64.7
	Agree to a greater extend	30	20.0	84.7
	Completely Agree	23	15.3	100.0
	Total	150	100.0	

Figure-7.25

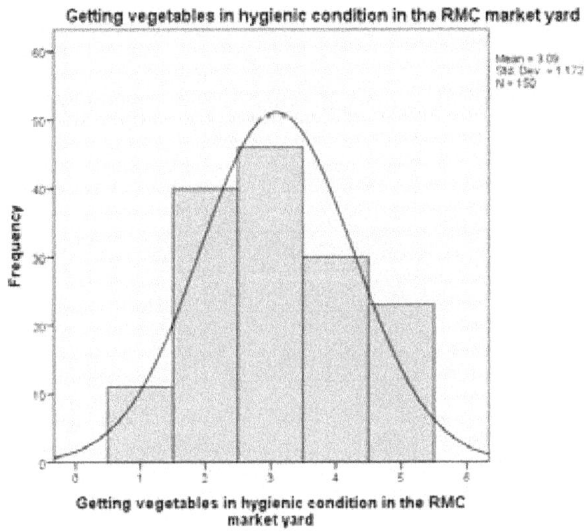

Table-28.26

Same quality vegetables are available through-out the day in this RMC market				
		Frequency	Percent	Cumulative Percent
Valid	Very Little Agree	10	6.7	6.7
	Little Agree	37	24.7	31.3
	Moderately Agree	59	39.3	70.7
	Agree to a greater extend	24	16.0	86.7
	Completely Agree	20	13.3	100.0
	Total	150	100.0	

Figure-7.26

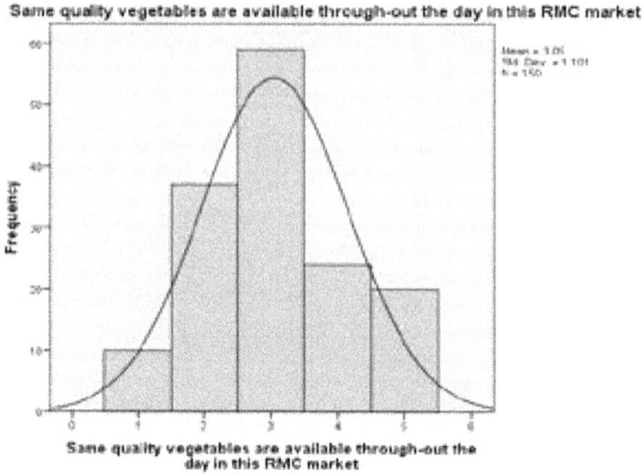

Table-28.2

Special carry bag for vegetable are provided by the seller of vegetables in RMC market				
		Frequency	Percent	Cumulative Percent
Valid	Very Little Agree	27	18.0	18.0
	Little Agree	38	25.3	43.3
	Moderately Agree	49	32.7	76.0
	Agree to a greater extend	20	13.3	89.3
	Completely Agree	16	10.7	100.0
	Total	150	100.0	

Figure-7.27

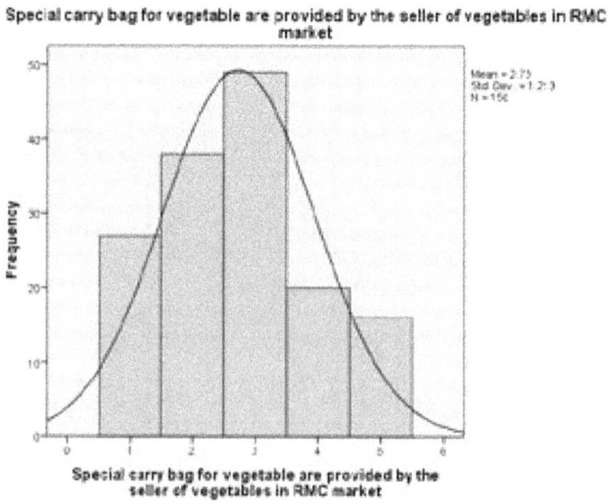

Table-28.28

Generally bargain for price with the seller while purchasing the vegetable products in this market				
		Frequency	Percent	Cumulative Percent
Valid	Very Little Agree	10	6.7	6.7
	Little Agree	21	14.0	20.7
	Moderately Agree	54	36.0	56.7
	Agree to a greater extend	32	21.3	78.0
	Completely Agree	33	22.0	100.0
	Total	150	100.0	

Figure-7.28

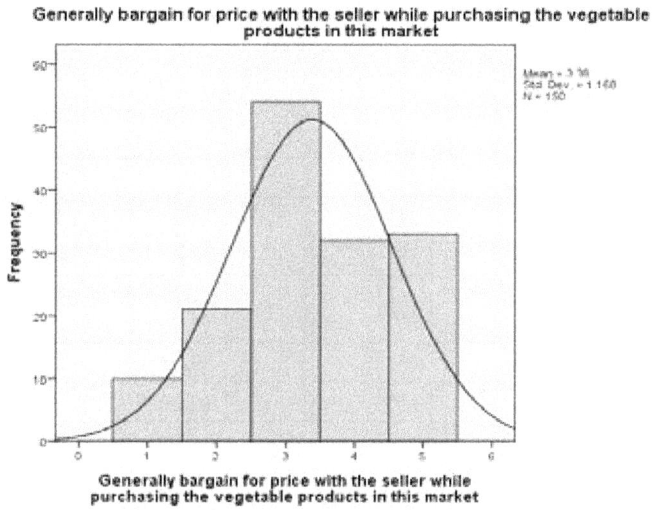

Table-28.29

Having more choices of sellers in this RMC market				
		Frequency	Percent	Cumulative Percent
Valid	Little Agree	20	13.3	13.3
	Moderately Agree	56	37.3	50.7
	Agree to a greater extend	26	17.3	68.0
	Completely Agree	48	32.0	100.0
	Total	150	100.0	

Figure-7.29

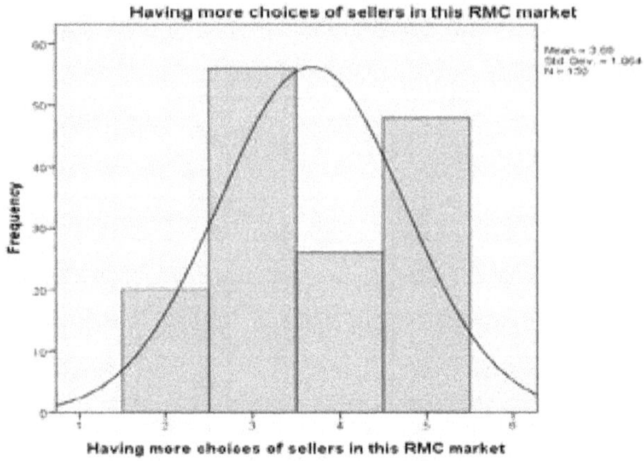

Table-28.30

Seeing almost the same sellers every day in this market				
		Frequency	Percent	Cumulative Percent
Valid	Very Little Agree	3	2.0	2.0
	Little Agree	21	14.0	16.0
	Moderately Agree	39	26.0	42.0
	Agree to a greater extend	42	28.0	70.0
	Completely Agree	45	30.0	100.0
	Total	150	100.0	

Figure-7.30

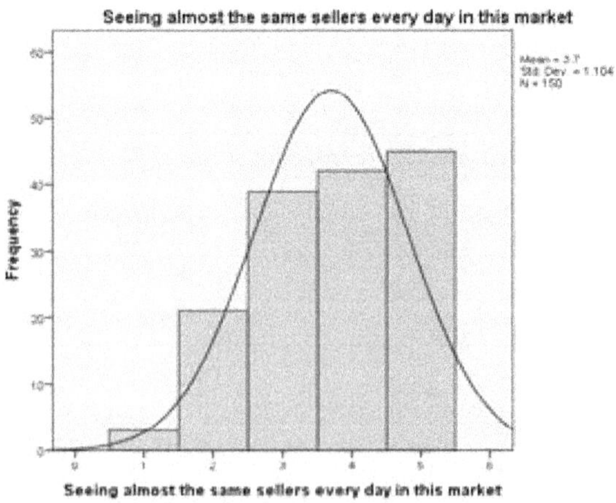

Table-28.31

Getting directly the firms products from the farmers-cum-sellers which can be store for several days				
		Frequency	Percent	Cumulative Percent
Valid	Very Little Agree	14	9.3	9.3
	Little Agree	25	16.7	26.0
	Moderately Agree	49	32.7	58.7
	Agree to a greater extend	31	20.7	79.3
	Completely Agree	31	20.7	100.0
	Total	150	100.0	

Figure-7.31

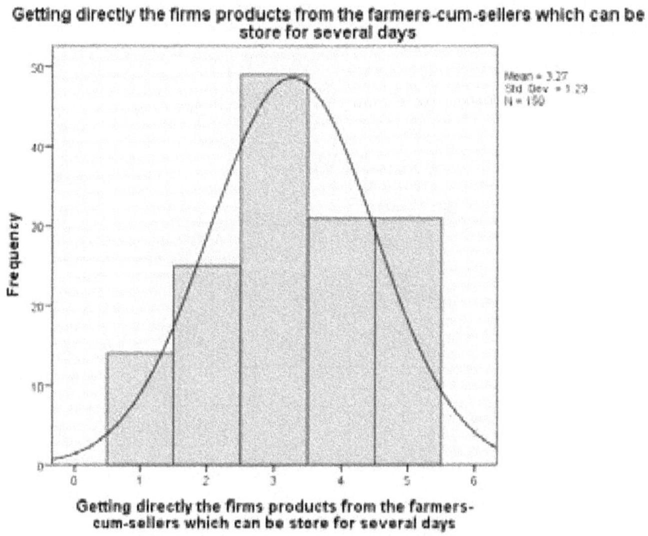

Table-28.32

RMC market yard have good parking place				
		Frequency	Percent	Cumulative Percent
Valid	Very Little Agree	22	14.7	14.7
	Little Agree	42	28.0	42.7
	Moderately Agree	50	33.3	76.0
	Agree to a greater extend	21	14.0	90.0
	Completely Agree	15	10.0	100.0
	Total	150	100.0	

Figure-7.32

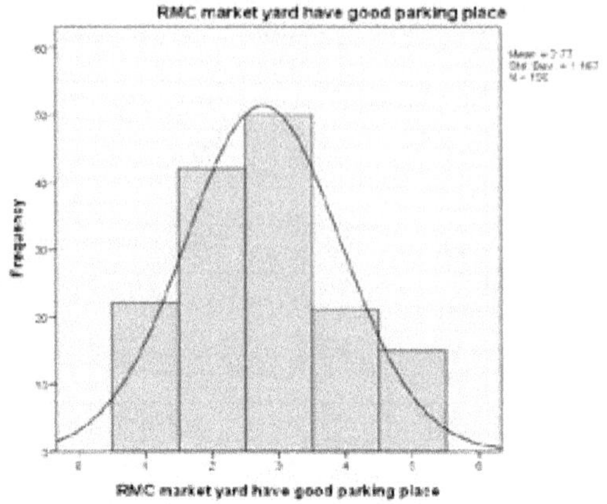

Table-28.33

RMC market yard is well structured for consumer facility				
		Frequency	Percent	Cumulative Percent
Valid	Very Little Agree	23	15.3	15.3
	Little Agree	29	19.3	34.7
	Moderately Agree	57	38.0	72.7
	Agree to a greater extend	23	15.3	88.0
	Completely Agree	18	12.0	100.0
	Total	150	100.0	

Figure-7.33

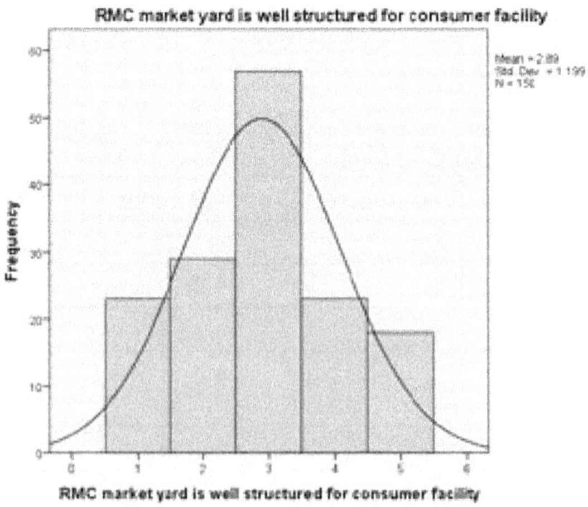

Table-28.34

Weight measurement in the RMC market seller is quite good				
		Frequency	Percent	Cumulative Percent
Valid	Very Little Agree	10	6.7	6.7
	Little Agree	29	19.3	26.0
	Moderately Agree	60	40.0	66.0
	Agree to a greater extend	26	17.3	83.3
	Completely Agree	25	16.7	100.0
	Total	150	100.0	

Figure-7.34

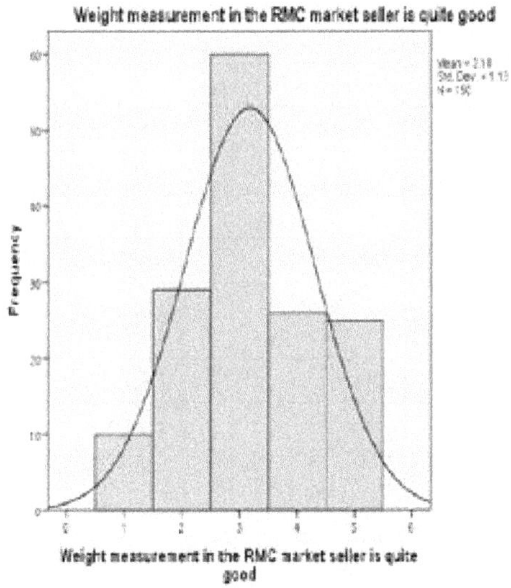

Table-28.35

Takes little more time to purchase vegetables from this market				
		Frequency	Percent	Cumulative Percent
Valid	Very Little Agree	5	3.3	3.3
	Little Agree	21	14.0	17.3
	Moderately Agree	62	41.3	58.7
	Agree to a greater extend	39	26.0	84.7
	Completely Agree	23	15.3	100.0
	Total	150	100.0	

Figure-7.35

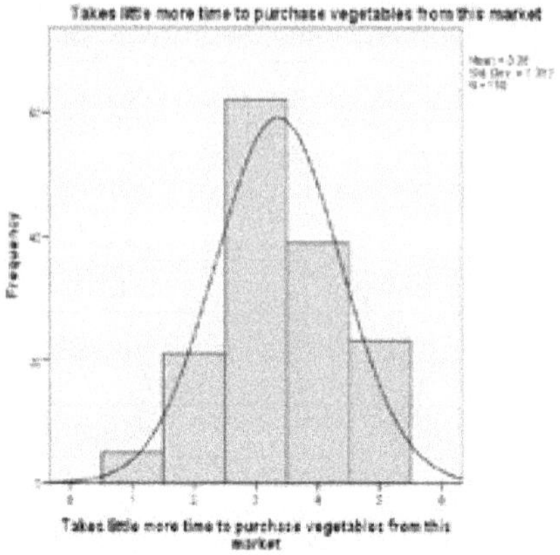

Table-28.36

Getting my product with in my budget prediction				
		Frequency	Percent	Cumulative Percent
Valid	Very Little Agree	6	4.0	4.0
	Little Agree	23	15.3	19.3
	Moderately Agree	58	38.7	58.0
	Agree to a greater extend	35	23.3	81.3
	Completely Agree	28	18.7	100.0
	Total	150	100.0	

Figure-7.36

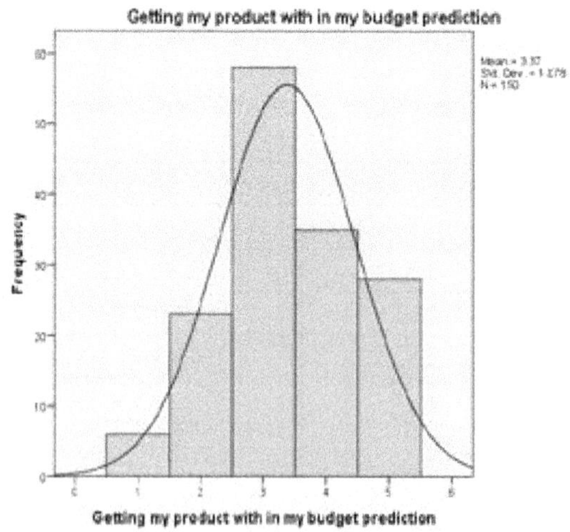

Table-28.37

I face problem to find out the organic vegetable in RMC market				
		Frequency	Percent	Cumulative Percent
Valid	Very Little Agree	3	2.0	2.0
	Little Agree	26	17.3	19.3
	Moderately Agree	48	32.0	51.3
	Agree to a greater extend	32	21.3	72.7
	Completely Agree	41	27.3	100.0
	Total	150	100.0	

Figure-7.37

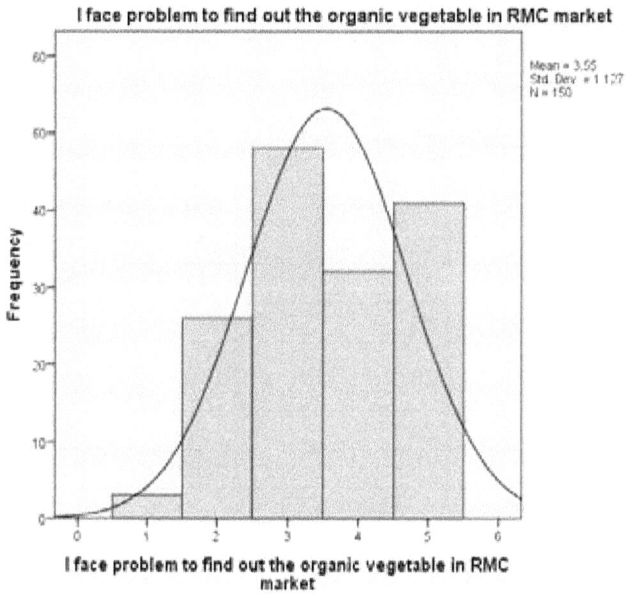

Table-28.38

I continue to purchase vegetable from the regulated market				
		Frequency	Percent	Cumulative Percent
Valid	Very Little Agree	8	5.3	5.3
	Little Agree	13	8.7	14.0
	Moderately Agree	66	44.0	58.0
	Agree to a greater extend	34	22.7	80.7
	Completely Agree	29	19.3	100.0
	Total	150	100.0	

Figure-7.38

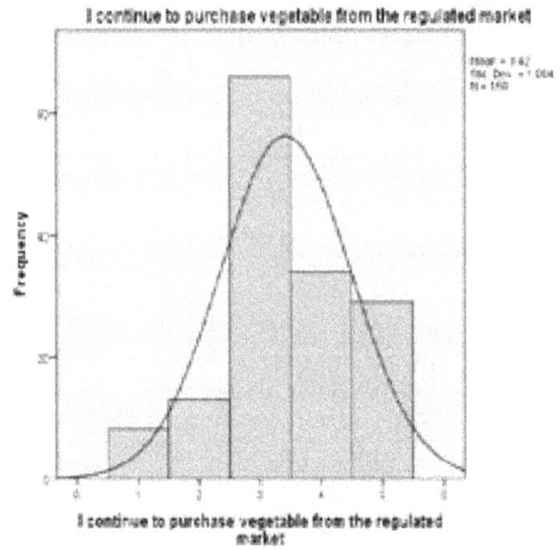

Table-28.39

This market has more space for shopping activities				
		Frequency	Percent	Cumulative Percent
Valid	Very Little Agree	12	8.0	8.0
	Little Agree	34	22.7	30.7
	Moderately Agree	55	36.7	67.3
	Agree to a greater extend	29	19.3	86.7
	Completely Agree	20	13.3	100.0
	Total	150	100.0	

Figure-7.39

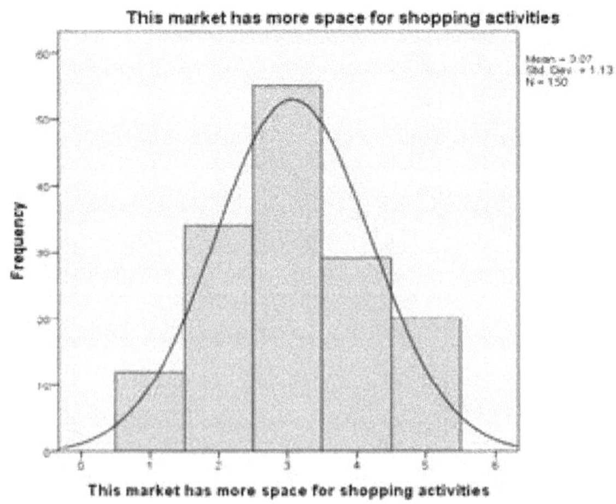

Table-28.34

Feel more rush in this market in holidays in comparison to other days.				
		Frequency	Percent	Cumulative Percent
Valid	Very Little Agree	2	1.3	1.3
	Little Agree	21	14.0	15.3
	Moderately Agree	44	29.3	44.7
	Agree to a greater extend	30	20.0	64.7
	Completely Agree	53	35.3	100.0
	Total	150	100.0	

Figure-7.40

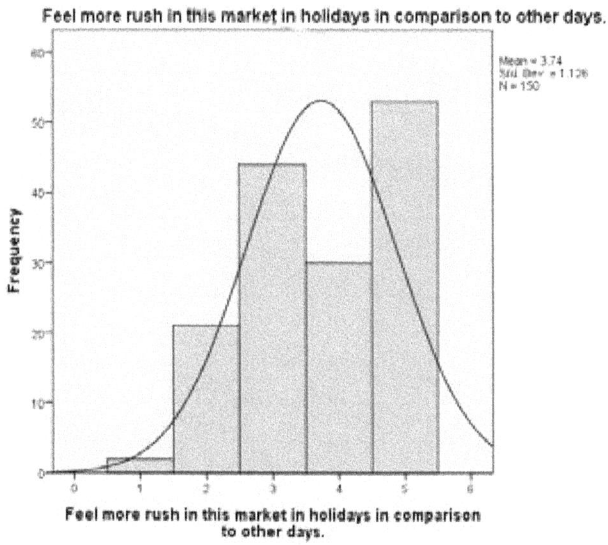

Table-28.41

Market place is well structured for the shopping & marketing activities				
		Frequency	Percent	Cumulative Percent
Valid	Very Little Agree	15	10.0	10.0
	Little Agree	31	20.7	30.7
	Moderately Agree	55	36.7	67.3
	Agree to a greater extend	26	17.3	84.7
	Completely Agree	23	15.3	100.0
	Total	150	100.0	

Figure-7.41

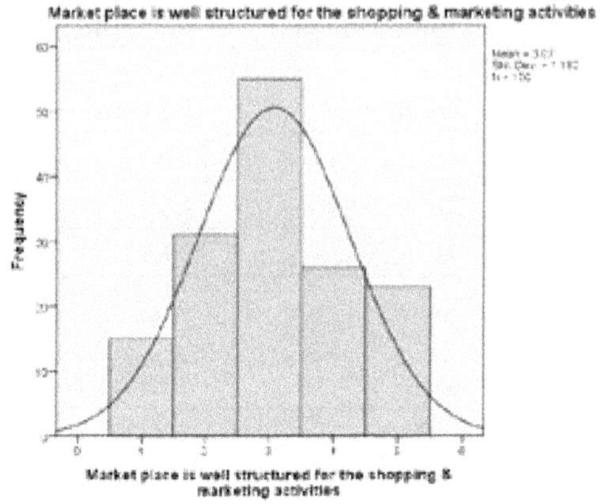

Table-28.42

Think farmers-cum-sellers are doing their business happily in this RMC market				
		Frequency	Percent	Cumulative Percent
Valid	Very Little Agree	12	8.0	8.0
	Little Agree	21	14.0	22.0
	Moderately Agree	53	35.3	57.3
	Agree to a greater extend	33	22.0	79.3
	Completely Agree	31	20.7	100.0
	Total	150	100.0	

Figure-7.42

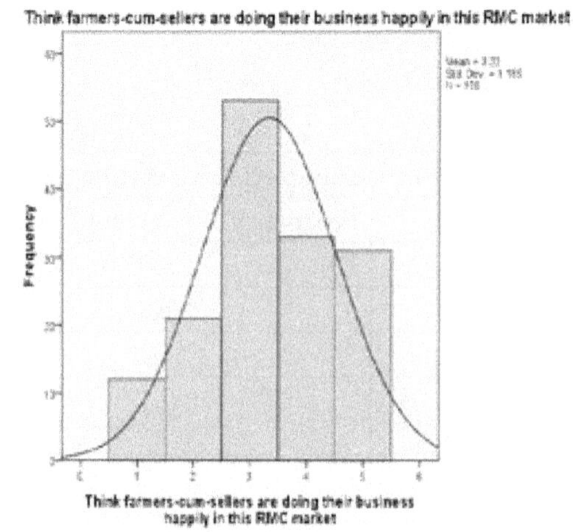

Table-28.43

Age				
		Frequency	Percent	Cumulative Percent
Valid	Below 20	5	3.3	3.3
	21-30	61	40.7	44.0
	31-40	62	41.3	85.3
	41-50	10	6.7	92.0
	Above 50	12	8.0	100.0
	Total	150	100.0	

Figure-7.43

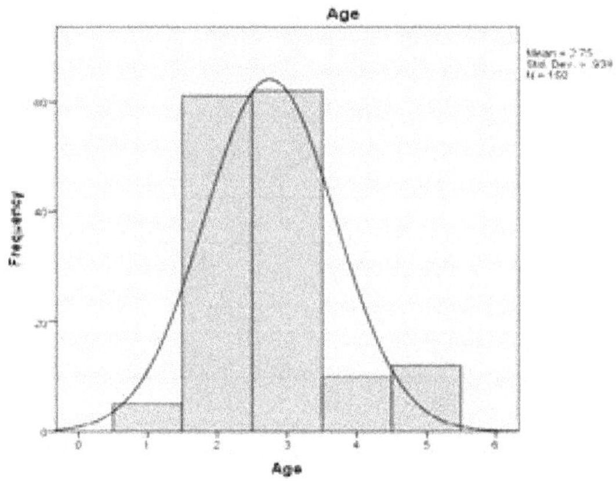

Table-28.44

Gender				
		Frequency	Percent	Cumulative Percent
Valid	male	102	68.0	68.0
	Femail	48	32.0	100.0
	Total	150	100.0	

Figure-7.44

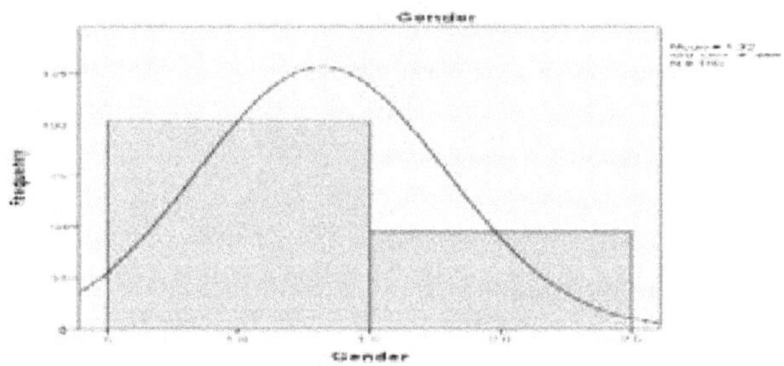

Table-28.45

Marital Status				
		Frequency	Percent	Cumulative Percent
Valid	Married	95	63.3	63.3
	Unmarried	55	36.7	100.0
	Total	150	100.0	

Figure-7.45

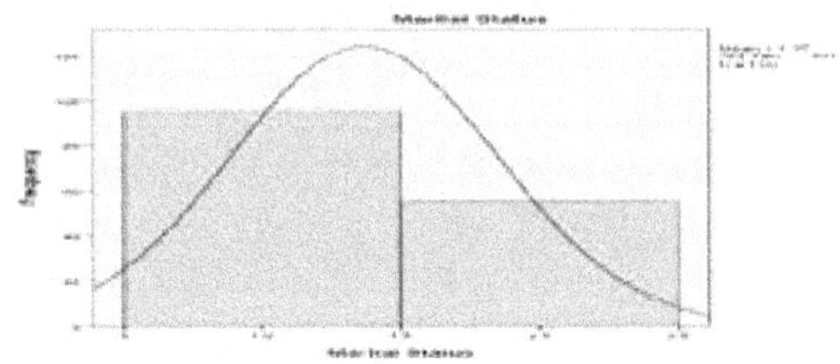

Table-28.46

Qualification				
		Frequency	Percent	Cumulative Percent
Valid	Bellow matriculation	6	4.0	4.0
	+2 level	24	16.0	20.0
	+3 level	56	37.3	57.3
	P.G level	64	42.7	100.0
	Total	150	100.0	

Figure-7.46

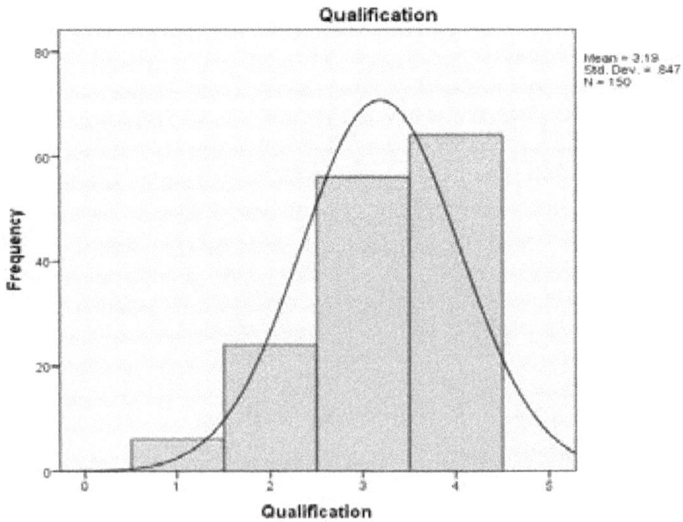

Index

Index